Geography
CHALLENGE!

A CLASSROOM QUIZ GAME

SECOND EDITION

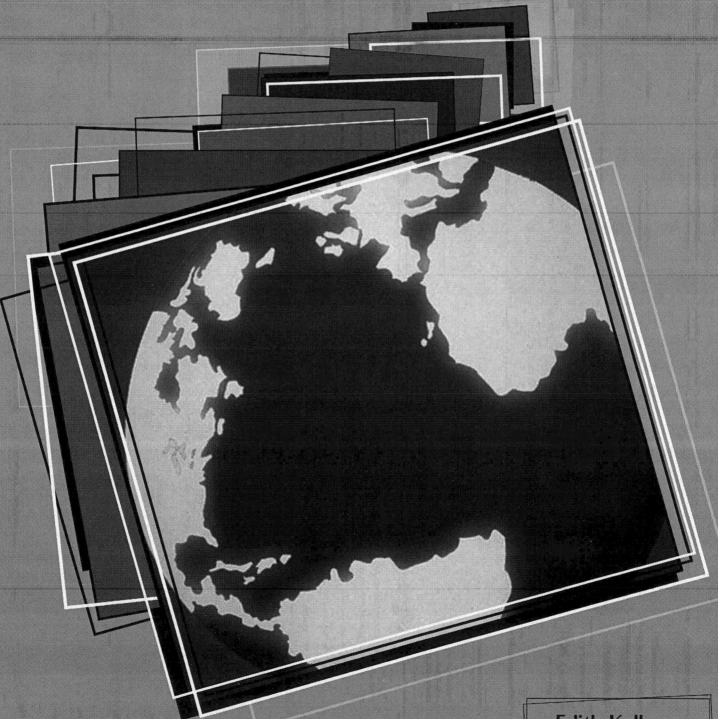

WALCH PUBLISHING

Edith Kellogg

SGS-SFI/COC-US09/5501

1   2   3   4   5   6   7   8   9   10

ISBN 0-8251-4359-4

Copyright © 1983, 1995, 2002
J. Weston Walch, Publisher
P.O. Box 658 • Portland, Maine 04104-0658
walch.com

Printed in the United States of America

Contents

Introduction

"The ... geographer looks at everything in the physical and human worlds because he realizes that everything is interconnected, directly or indirectly, and changing at all times. Around the globe is a constantly moving mass of gasses, water vapor, trace elements, ice, pollution, people, ideas, money. This moving mass is driven essentially by the energy of the Sun and humankind."

—*George J. Demko, author of* Why in the World

Geography so touches our daily lives! How could the subject of world geography have been dismissed as unimportant? Now our woeful ignorance and lack of curiosity about the world beyond our borders has become a liability. Today, in our shrinking world of expanded communication and interdependence, the study of world geography is more essential than ever before.

Geography is a correlated science, drawing together two classes of interrelated elements: the diverse physical features of the many areas of our earth and the cultural influence of human occupancy. People are the central themes—where they live, their stage of development, and how they adapt to, utilize, and change their habitats.

An insight into the lives of people beyond a student's own area will help bring about not only the student's appraisal of his or her own community's advantages and limitations, but also a sympathetic understanding and greater appreciation of the potentialities and problems of these other cultures.

We need this awareness of the interrelationship of people and their earth. We must know the earth before we can know how to protect its resources, how to foster worldwide economic development, and how to live in harmony with it—both ecologically and humanly. Our future leaders must depend upon this earth to support their generation and future generations.

To the Teacher

Geography Challenge! deals with facts and under-standings about all regions of the earth. It is designed to be used for several purposes: as a fun and easy way to reinforce what is being studied, as a study guide, and as a review of the unit or a culminating activity. It challenges your students to remember important facts and encourages them to enjoy themselves in the process.

The format of *Geography Challenge!* is similar to that of a popular television game show. A student is given the answer and is asked to provide the question. The fact given as a question is actually stated ("Land adjoining a sea or ocean"), not asked. The student response is given as a question ("What is a coast?"). Many students will already be familiar with the format.

The questions are classified according to general topic and further by category. This format lends itself to use with a variety of attention-keeping games. Some games are suggested here; you and your students may invent others.

A number of questions throughout the book have more than one correct response. Often, an alternate response is written in parentheses, for example, "What is the greenhouse effect (or global warming)?" Sometimes a whole answer is given but a less complete one would be acceptable. For instance, in reference to Poland's coalfields, the answer is written: "What is (Upper and Lower) Silesia?" The parentheses indicate that Upper and Lower may be omitted.

Population figures and comparisons refer to metropolitan areas, which include the central city, built-up suburbs, and more distant communities if the bulk of the residents are employed in the central city. The United States figures are based on 1998 or 1999 population estimates, and those of the rest of the world are taken from mid-2000 estimates.

Many place names have both an anglicized and a native version. Some of the native names are becoming commonly accepted. When both are used, the anglicized word is written first, with the native name following in parentheses, for example, Yangtze River (Chang Jiang).

How to Use This Book

Unit 1 is a study of basic geographical terms and concepts, dealing with landforms, climate and vegetation, and maps. Unit 13 is a summarizing chapter that includes terms, situations, and concepts not applicable to any one particular region. The other units are set up according to world cultural groupings as they might be studied; the United States is further divided into regional topics.

Each topic, or game, consists of five general categories. Within each category are five questions, each assigned a point value of 5 through 25 depending on its relative difficulty, plus a bonus question. The bonus question is not necessarily more difficult; it may refer to an unusual fact or a less important one. It may be used in whatever way seems suitable. A point value of 5 for each bonus question would give the entire game 400 points; a value of 25 would make it a 500-point game.

These questions in this format may be used to play a variety of games. However, it may prove effective to allow the students an opportunity to find the answers to, or study, the questions first. You may wish to reproduce the questions for a series of assignments, and then use a game as an evaluation, a further review, or a culmination of the unit. You may find that using the questions without a game is adequate. For these reasons, the answers are presented separately at the back of the book rather than with the questions.

Feel free to modify *Geography Challenge!* If you have stressed something in your class that is not included in this game, it is easy to add questions. Your students will quickly learn how to make questions for you in order to extend the game. You can also modify the questions to make them easier or harder to fit the needs of each particular class. Your class can play the same game more than once, which will help them remember material much more easily.

The same basic procedure can be used for playing any number of different games. Here are the directions for a typical game:

- Write on the board the categories for the game to be played along with point values for each question.

- Divide the class into teams. Play begins when one student asks for a question from a given category with a given point value. For instance, if the topic is Physical Features of South America, the student might say, "I want the 10-point question from the 'Climate' category."

- The game leader then reads the 10-point question from the requested category.

- Any student on the team may answer. The first person on the team to raise his or her hand is called on. (It may be the student who asked for the category to begin with.)

- If the answer is correct, record points for the team. The student who answered chooses the category and point value for the next question.

- If the answer is wrong, subtract the point value of the question from the team score. A student from the other team now has the chance to answer the question. Whoever answers the question correctly chooses the category and point value for the next question.

- If no one can answer the question, give the correct answer to the group. The student who last successfully answered a question chooses the next category and point value.

- When all the questions in the category have been used, erase the category from the board. Continue until all the categories are erased and the game is over.

Following are some other variations of the game:

Rounds

The categories and point values are displayed and the value of the bonus question is agreed upon. Bonus questions are not used until last. A scoreboard is drawn on the board to show the teams and what score they receive in each round.

The class is divided into three, four, or five groups, each having an equal number of students. (Up to 30 can play. Extra pupils may serve as scorekeepers, readers, or board keepers.) The players in each group or team sit or stand in a set order—first player, second, etc.

The game begins with Player 1 on Team 1 requesting a question. If the player responds correctly, the earned score is recorded under Team 1/Round 1. If the response is incorrect, the correct answer is read and a score of 0 is recorded. In either case, the point value is erased under the respective category. Then Player 1 of Team 2 has a turn to choose a question. After all the first players on each team have played, the play goes to the second players of each team, then the third, and so forth.

The game continues for as many complete rounds as possible. There may be several unused questions. If there are 30 players, the last player in each team chooses a category for a bonus question. Otherwise, the bonus question for each team is given to, or chosen by, the team's top scorer or chosen captain, either for that player or for the team to answer. The top-scoring team wins.

Progression

This game is set up like Rounds, preferably in five groups. The first players on each team choose a category for 5 points, the second players choose a question for 10 points, the third players go for 15, etc. Play continues for as many complete rounds as possible, with bonus questions handled as in Rounds.

Concentration

First, the categories and point values are written on the board and the bonus value is determined. The class is divided into two teams. The first player on one team requests a question. If the player replies correctly, his or her team gets the points, and the point value is erased below the respective category. If the player does not answer correctly, the response is announced to be wrong and nothing is erased from the board. The first person on the opposite team then chooses a question. The play goes from team to team, with each person choosing a question still listed on the board. The advantage goes to the person who knows the answer to a previously asked question and can remember where it is located on the board. Play continues until all questions have been used. The highest-scoring team wins.

Last Chance

The class is divided into two, three, four, or five teams, with the players seated or standing in a set order. The categories and point values are displayed, and the bonus value (perhaps generous) is chosen. The bonus questions are not used in regular play.

Player 1 on the first team requests a question. If the player replies correctly, his or her team earns the respective points; if the reply is incorrect, the teacher tells or explains the answer. In either case, the point value under that category is erased. The play then goes to Player 1 on the second team, who requests a question. After all the first players have had a turn, the play goes to the second players on each team, then the third, and so forth.

When all the questions have been used, the scores for each team are calculated. The next player on the lowest-scoring team chooses a category for the bonus question for his or her team. The teacher reads the question and accepts only one answer from the team.

(The players may confer in order to come to an agreement.) If the reply is correct, the bonus score is added to their total. Then the second-lowest-scoring team chooses a category, then the third, and the fourth, if there are that many teams. Only one bonus question is given to each team. There may be some that are not used. The winning team is that which has the highest score.

Solo

This game is played like Last Chance, except that it is played by five players instead of teams. The play goes from one player to the next in succession until all questions are used. Then each has a chance to choose a bonus question to raise his or her score. The top scorer wins.

Geography Bee

This game is played like a spelling bee, but no one is eliminated. First the categories and point values are displayed, and the value of the bonus question is determined. The class is divided into two teams. The first person on one team asks for a question by stating a category and point value. If the player responds correctly, his or her team receives the points and that point value is erased under that category. The next turn is taken by the first player on the other team, who chooses a question. However, if the first player's response is not correct, the same question is repeated for the first player on the other team. If the player replies correctly, his or her team gets the points and the play then goes to the second player of the first team. The play continues from one side to the other, with points going to the teams that answer correctly and the respective category points being erased from the board. The game is over when all 30 questions have been used. The team accumulating the most points wins.

No matter how you use *Geography Challenge!* it is an entertaining and stimulating way to review, and it's an excellent change-of-pace activity. You'll find your students eager to play it again and again.

UNIT 1

Geographical Terms and Understandings

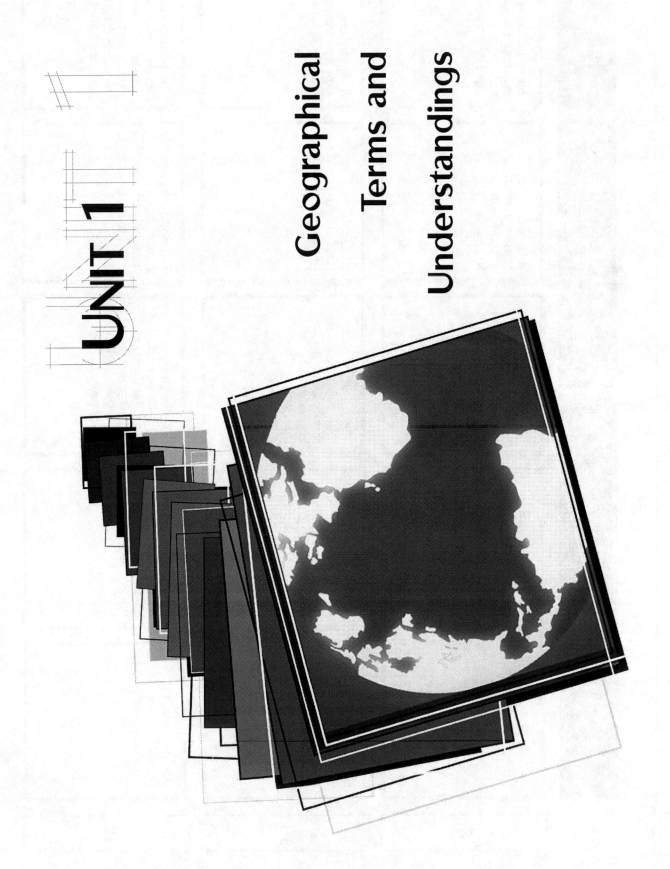

Geographical Terms and Understandings

Topic: Maps

1

	BASIC TERMS	LATITUDE	LONGITUDE	KINDS OF MAPS	MAP READING
5	Half of the globe	An imaginary circle around the middle of the earth, halfway between the North Pole and the South Pole	An imaginary rod on which the earth rotates	A true representation of the shape of the earth	A key telling what symbols are used on a map to represent various features
10	Distance east or west of the prime meridian, measured in degrees	The parallel of latitude that is 23°30' north of the equator	The meridian of 0° longitude that passes through Greenwich, England, and from which other longitudes are measured east and west	A map that shows the natural features of the earth's surface	A letter, picture, or sign that represents something on a map
15	Distance north or south of the equator, measured in degrees	The parallel of latitude that is 23°30' south of the equator	An imaginary line where each calendar day first begins, running mostly along 180° longitude	A map that shows boundaries of nations and other cultural features	The relationship between a given distance on the earth's surface and the same distance as represented on a map

A system comprised of parallels and meridians for locating places on the earth's surface

Lines drawn on a map connecting all points that are the same altitude above the sea

Lines on a map that join all places having the same temperature

A map that shows elevations as well as the positions of points, often in color and with contour lines

A map that employs different colors, in a definite order, to distinguish land at differing heights and water at varying depths

A cylindrical map projection in which the distortion of areas increases as it nears the poles, but is essential for marine navigation

A series of 24 zones, each 15°, determined approximately by meridians, within which the same standard time is used

The direction from the international date line that the next day begins

A system instituted by the United States to take advantage of the extra hours of daylight in the summer

The imaginary boundary of the north polar region, at 66°30' north latitude

The imaginary boundary of the south polar region, at 66°30' south latitude

Either end of the earth's axis, located at 90° north latitude or 90° south latitude

An imaginary line that passes through all points of the same longitude

An imaginary line on the earth connecting points of equal latitude

The making and study of maps and charts

20

25

B O N U S

N O T E S

Geographical Terms and Understandings

1

Topic: Land

	BASIC TERMS	LANDFORMS	ELEVATIONS	CHANGE PROCESSES	POTPOURRI
5	One of the seven great divisions of land on the earth	A body of land entirely surrounded by water	A natural elevation of the earth's surface, rising high above the surrounding level	A trembling of the earth's surface, caused by changes beneath the surface	The shallow portion of the seabed bordering most continents and terminating in a slope, which descends steeply into deeper water
10	The level of the ocean, which is the same all over the world	A narrow portion of land nearly surrounded by water	A raised part of the earth's surface, smaller than a mountain	A mountain or hill having an opening through which molten rock, steam, and ashes are expelled from within	A mountain ridge across a continent, which separates streams flowing in opposite directions
15	The elevation of a place above sea level	A point of land jutting out into the water	A long depression or hollow lying between hills and mountains	A break in the earth's crust, where the land on either side has moved up, down, or sideways along the break	The points at which the land drops sharply and rivers usually become unnavigable

20

- Land adjoining a sea or an ocean
- A narrow strip of land connecting two larger portions of land
- An area of mostly high, generally level land
- A long dry period caused by the absence of rain
- A hollow or depression in the earth's surface, almost entirely surrounded by highlands

25

- A large part of a continent that forms a section of its own
- A group of many islands
- A broad area of flat or gently rolling land
- The act of making lands suitable for human use
- A chain of mountains

BONUS

- Not having an outlet to an ocean or a sea
- A ring-shaped coral reef island that surrounds a lagoon
- A deep, narrow valley with steep sides
- Tiny sea animals whose skeletal remains can build up to form a reef
- A ridge of rock, sand, or coral at or near the surface of the water

NOTES

Geographical Terms and Understandings

1

Topic: Water

	BASIC TERMS	WATER FORMATIONS	RIVERS	WATER ON LAND	POTPOURRI
5	One of the four largest bodies of water on the earth	A body of freshwater or salt water of considerable size, surrounded by land	The beginning of a brook or river	A large river of ice formed from snow on high ground	An artificial waterway for navigation or irrigation
10	A large natural stream of water	A portion of the sea extending into the land	The part of a river where its waters empty into the sea or into another river	An area where the soil is under shallow water, such as a swamp, marsh, or bog	A place where water is collected and stored for future use, especially an artificial basin created by the damming of a river
15	The movement of water from the earth to air and back again to the earth by evaporation, condensation, and precipitation	Part of a body of water that reaches into the land, smaller than a gulf	A river that flows into a larger river	Water that lies beneath the surface of the earth	A part of a river's course where the water rushes over rocks

Geography Challenge!

©1983, 1995, 2002 Walch Publishing

20	25	BONUS	NOTES
A body of water encircled by a coral reef	A very powerful ocean wave, set off by earthquakes or undersea volcano eruptions, that causes great destruction on land	A floating mass of ice that has broken off from a glacier	
A usually dry river channel, found in desert regions, that fills with running water during a cloudburst	A bed of a salt lake that contains water at irregular periods	The level below which the ground is saturated with water	
Land built up by deposits at the mouth of a river	The broad mouth of a river into which the tide flows	A riverbank higher than the surrounding land	
A narrow waterway connecting two large bodies of water	A narrow inlet of the sea between the steep sides of a glaciated valley	A narrow strip of water extending into a coast	
The flow of a stream of water, either in a river or through an ocean	The regular rising and falling of the ocean	The study of the ocean and the life within it	

Geographical Terms and Understandings

Topic: Climate and Weather

1

BASIC TERMS	CLIMATE TYPES	WINDS	OCEAN CURRENTS	STORMS
5 The condition of the atmosphere at one point on the earth's surface at a given time	The type of climate characterized by sparse rainfall, with hot or variable temperatures	Pertaining to the side facing the wind	A warm ocean current issuing from the Gulf of Mexico and flowing northward along the eastern coast of the United States	The most common storm worldwide, characterized by lightning, thunder, and, usually, heavy rain
10 The average conditions of weather in a region over a period of years	Term for a hot climate, with either heavy rainfall year-round or wet summers and dry winters	Pertaining to the side away from the wind	A cold ocean current that rises in the Arctic Ocean and follows the eastern coast of Canada southward	A funnel-shaped cloud with violent winds that affects a limited area
15 The layers of air that surround the earth	Term for a climate typical of the interior of continents, characterized by hot summers and cold winters, with varying amounts of precipitation	The usual westerly winds that blow between 30° and 60° latitude, both north and south of the equator	The continuation of the Gulf Stream that carries warm water northeastward past the northwest coast of Europe	A tropical late-summer storm accompanied by high winds and heavy rain

©1983, 1995, 2002 Walch Publishing

A seasonal wind of southern Asia that brings heavy rain from the south in summer and dry air from the north in winter

A tropical storm in Southeast Asia accompanied by high winds and torrential rain

A narrow band where two contrasting air masses meet, interact, and often produce stormy weather

A cold ocean current in the Pacific Ocean that flows northward along the west coast of South America

The warm current that flows from the Pacific tropics along the east coast of Japan

The current that carries warm water southward from the tropics along the eastern coast of South America

Winds blowing steadily over the ocean toward the equator, from the northeast in the northern hemisphere and the southeast in the southern

Certain regions near the equator where the wind is very light or constantly shifting

The weight of air on the surface of the earth, which varies from place to place

Term for a climate of the lower latitudes, where the summers are hot and dry and the winters are warm and rainy

Term for a climate influenced by the sea, having warm summers, cooler winters, and adequate rainfall year-round

Term for the climate type characterized by cool summers, cold winters, and light precipitation

The depositing of moisture in the form of rain, hail, snow, sleet, or ice on the earth's surface

The amount of moisture in the air

The study of atmospheric conditions as they pertain to weather

20

25

BONUS

NOTES

Geographical Terms and Understandings

Topic: Soil and Vegetation

1

	BASIC TERMS	SOIL	VEGETATION REGIONS	CHANGE PROCESSES	PLANT POTPOURRI
5	All the surrounding conditions and influences that affect the development of a living thing	Decayed organic plant material	The natural vegetation of equatorial areas having a constant heavy rainfall	A gradual wearing away by glaciers, temperature changes, running water, waves, ice, and wind	The period from the last frost in spring to the first frost in fall during which crops can be raised
10	Careful saving and protection of something, such as natural resources	Permanently frozen soil	An area of sparse precipitation whose vegetation is limited to scrub, cactus, and grasses	Dissolving of soil minerals by rainwater that carries them down beneath the root zone	The boundary north of or above which trees cannot grow
15	A small area in a desert region that has water and fertile soil	Pertaining to soil well suited for growing crops	An area where cool summer temperature and light precipitation can only support mosses, lichens, and grass	Adding natural or chemical nutrients to the soil to make it more productive	Pertaining to trees or plants that remain green the year-round

20			
Condition of receiving little rain because of being on the protected side of high mountains	Relating to soil formed from the material gradually deposited by flowing water, such as clay, silt, or sand	A tropical land with tall grasses and scattered trees that has wet summers and dry winters	Supplying dry land with water in order to make it produce crops	Pertaining to trees or plants that shed their leaves during one season of the year

25			
The study of the relationships among living things and their surroundings	Fine earth materials deposited by running water or wind	A dry, level, short-grass region in the midlatitudes	Level land built up by deposits from a river that overruns its banks	Temporarily inactive

B O N U S			
The science that deals with the earth's crust, the layers of which it is composed, and their history	Rich, fertile earth in which decaying leaves and silt are mixed with clay and sand	A needle-leaf forest of the dry subarctic regions	The breakdown of surface rocks by physical or chemical forces	Stunted trees and bushes growing in poor soil or in semiarid regions

N O T E S

UNIT 2

The United States

The United States
Topic: Northeastern States

	LAND AND WATER	CLIMATE AND RESOURCES	LAND USE	INDUSTRIES	PEOPLE AND PLACES
5	The largest bay on the Atlantic coast, actually an estuary, which divides Maryland into two sections	The area in which sandy soils, milder winters, and proximity to markets promote truck farming	The nation's largest inland port, situated at the junction of three rivers	The planned city located between Maryland and Virginia, but not a part of either state, which is the U.S. capital and whose business is government	The largest city in the United States and a national center of trading, manufacturing, communication, education, finance, and culture
10	The important river formed at Pittsburgh from the junction of the Allegheny and Monongahela rivers	The reason for the growth of early industries at the fall line and in New England	The type of farming suited to poorer soil, a short growing season, hilly land, and nearness to markets	The area of New York City that, because of its stock exchanges, many banks, and commercial offices, is the world's greatest financial center	The City of Brotherly Love, the region's second largest city, a great port and manufacturing center, which boasts a rich historical heritage
15	The boundary at which eastward-flowing rivers fall over the edge of the Piedmont onto the Atlantic Coastal Plain	Hard coal, found near Scranton and Wilkes-Barre in Pennsylvania, which heats well but is difficult to mine	The area which, together with the coastal waters off Cape Cod, accounts for a large share of the nation's catch of fish and shellfish	The commercial industry in which the states of New York and Connecticut lead the nation	A major port of New England, a leading industrial city for fishing, printing, and finance, and a major center of learning and health care

The name for a continuous urban area where one metropolitan area merges with the next

That which city dwellers take with them when they move into the suburbs, creating serious problems for the cities

An organization of world nations, with headquarters in New York City, that works for peace, cooperation, and development

The area in which traditional industries have declined the most but are being replaced by new and varied industries

Technical products manufactured in the Northeast, such as computers and scientific instruments

The Chemical Capital of the World because it serves as headquarters for many chemical companies and has many research laboratories

Once responsible for the growth of shipbuilding in the Northeast, the resource that is now used largely for pulp and paper

A major industry of New England, due to the area's snow-covered mountains, beautiful scenery, extensive coastline, and historic interest

The crop for which Maine's Aroostook Valley is famous

Waterfalls on the river connecting Lake Erie and Lake Ontario, which produce much hydroelectric power for the area

The tree that grows in the northern part of this region and yields a sap used for sugar, syrup, and candy

A hard rock used in fine architecture, which is quarried in Vermont, as are slate and granite

A plateau of rolling hills, crossed by rivers, that lies between the Atlantic Coastal Plain and the Appalachians

The river that separates the Green Mountains and the White Mountains, and in whose valley tobacco is grown

The naturally beautiful hilly area comprised of six states, which was one of the first parts of the United States to be settled and industrialized

20

25

B O N U S

N O T E S

The United States
Topic: North Central States

2

LAND AND WATER	CLIMATE AND RESOURCES	LAND USE	INDUSTRIES	PEOPLE AND PLACES
5 The level, sometimes rolling, gradually sloping region that stretches from the Appalachian Highlands to the Rocky Mountains	The type of wheat grown chiefly in the milder climate of Kansas, covering the topsoil most of the year and preventing erosion	The crop region extending from Ohio to Nebraska on which cattle and hogs are also raised	The state that produces most of the iron ore for the two main industries of the Great Lakes states—iron and steel	A major industrial city that is also a national transportation crossroads for railroad and air traffic and an inland and ocean port
10 The river that, with the Missouri and Ohio, drains one half of the United States into the Gulf of Mexico	A kind of storm that threatens farms on the Great Plains, brought on when strong winds follow a drought	The crop now grown in the drier Plains where once there were prairies and cattle ranges	A major industry of the Midwest that depends on the cattle and pigs that are raised and fattened there	Once the "auto-mobile capital of the world," this city now manufactures steel, pharmaceuticals, processed foods, machine tools, as well as automobiles and auto parts.
15 The only one of the five Great Lakes entirely in the United States, dividing Michigan into two parts	The spring and summer storm of the southern Great Plains in which violent whirling winds cause great destruction	The state in the Hay and Dairy Belt known as America's Dairyland because it leads in dairy cattle and milk products, especially cheese	A leading industry of Minneapolis and Kansas City in which wheat and other grains are processed	A metropolitan area on both the Kansas and Missouri sides of the Missouri River that is a leading livestock and wheat market and transportation center

20

- The main tributary of the Mississippi River, upon which dams were built for flood control, irrigation, and navigability
- The fossil fuel abundant in Illinois which is used to generate about half the state's electricity
- Bins in which wheat is stored until it is shipped to market
- One of the greatest concentrations of heavy industry in the United States, located in the vast urban area in Indiana that borders Lake Michigan
- The Gateway Arch to the West, a major transportation center due to its location, and a leader in the chemical, auto parts, and food-processing industries

25

- The waterway that links much of the Midwest, via the Great Lakes, with ocean trade routes
- The two leading corn states that have some of the most valuable farmland in the United States
- Another crop grown in the Corn Belt as feed for livestock and as a source for oil, flour, and other products
- A highly-industrialized city in northeastern Ohio, still an important center of rubber research and development
- The chief port on Lake Erie within easy reach of raw materials and markets that manufactures steel and steel products

BONUS

- Low mountains in South Dakota that eroded into gigantic rock stubs and became the site of the Mount Rushmore Memorial and a large gold mine
- The natural force that leveled the Plains area and left behind deep, rich, black soil
- Organizations of several small farmers in which the costs and profits of farm production are shared
- The Kansas city and its surrounding area known for the manufacture of aircraft and aircraft parts
- A lake port and industrial city specializing in farm machinery, engineering, and breweries

NOTES

2 The United States
Topic: Southern States

	LAND AND WATER	CLIMATE AND RESOURCES	LAND USE	INDUSTRIES	PEOPLE AND PLACES
5	The river that, with its main tributary the Missouri, forms the third longest river system in the world	The mineral found in abundance in the Appalachian Mountains, with West Virginia and Kentucky leading the United States in its production	The area extending from Florida to North Carolina and west to California, named for the crop used for textiles and valuable by-products	The industrial process for which Tulsa, Oklahoma, and Texas Gulf cities such as Houston and Galveston are noted	A famous southern city known for its French influence, and one of the world's busiest ports due to its location at the mouth of the Mississippi
10	The submerged part of the Coastal Plain where the waters abound with fish and where large amounts of oil and natural gas are found in the Gulf region	The kind of fruit grown in the hot, humid climate of Florida and southern Texas	A crop traditionally grown along the South's northern edge, especially in the Nashville and Bluegrass basins	A rapidly-growing food industry due to modern production techniques and health consciousness	The two neighboring cities that comprise the South's largest urban center, serving as a hub of finance, manufacturing, transportation, and communication
15	The region of low rolling hills southeast of the Appalachians, separated from the Coastal Plain by the fall line	A leading power source that can be piped directly to homes, with over 60 percent of the U.S. supply produced by the South	An abundant resource of the Appalachians and swampy coastal areas, used in the manufacture of pulp and paper	The term used for the various products having oil as a base ingredient	A world-famous resort city on Florida's east coast, one of many in this state of year-round sunshine and tourist attractions

One of the poorest areas of the United States because of its barren, hilly land and isolation, which is being helped by government programs

One of the fastest growing cities in the United States, whose location makes it a center of air and road routes as well as industry

A leading Texas port connected to the Gulf by a ship canal, and a leading center of the oil-refining and petrochemical industries

The location of the NASA space center and the principal U.S. launching site for Earth satellites and space flights

Once a steel-producing city, now a regional transportation, commercial, and cultural center whose largest employer is the University of Alabama

The type of coal used in blast furnaces, which is made in great quantity from bituminous coal

The type of agriculture that occupies much of the original cotton land and is helping to rebuild the depleted soil

A regional food crop, grown especially in Georgia, which is also used for oil, livestock feed, and manufactured products

An important government project begun in 1933 whose purpose is to control floods, develop river travel, and harness water for cheap electricity

Two minerals, both starting with the letter S, that are used in chemical production and are found in the coastal areas of Texas and Louisiana

The ore used for producing aluminum, its only U.S. deposits lying in Arkansas, Alabama, and Georgia

A shellfish caught in the Gulf waters of Louisiana, Texas, and Florida, which is frozen or canned for domestic and foreign markets

The name given to the hills in southern Missouri and Arkansas

The plateau in western Texas whose rich grasses feed sheep, goats, and beef cattle

Broad, strong banks of earth and concrete that border rivers of the Mississippi system to hold back floodwaters

20

25

B O N U S

N O T E S

Geography Challenge! 19

The United States

Topic: Western States

2

LAND AND WATER	CLIMATE AND RESOURCES	LAND USE	INDUSTRIES	PEOPLE AND PLACES
5 The boundary where the high, rugged Rocky Mountains separate the rivers flowing east from those flowing west	The main river of the Southwest whose available water supply was divided among Arizona, California, and Nevada by a decision of the U.S. Supreme Court	The process that allows the West's fertile, dry lands to be farmed, compensating for the few rivers and inadequate rainfall	The most populous state, rich in natural resources, which grows almost every crop, and produces oil, transportation equipment, and processed foods	An enormous metropolitan area, the West's chief business and trade center, and the leader in the aerospace and motion-picture industries
10 The valley formed by the San Joaquin and Sacramento rivers, bordered on the east by the Sierra Nevada and on the west by the Coast Ranges	The northwestern river, shared with Canada, whose many dams are a major source of hydroelectric power and irrigation	The dominant industry of the Pacific Northwest, dependent on the area's abundant rainfall and marine climate	The newest state, which attracts an important tourist trade with its climate, volcanoes, customs, and exotic scenery	The beautiful City by the Golden Gate, a port trading with the Far East, and a financial, industrial, and cultural center
15 The Hawaiian island with the city that serves as the state's capital, largest city, and chief seaport	Rock from which crude oil can be distilled, with Wyoming and Colorado having the largest reserves	The prime catch of the Alaskan and Pacific Northwest's fishing industry	The state that is home to many government installations and that employs many in research and development of nuclear energy and exploration	A large port on Puget Sound, and an industrial city producing aircraft, ships, lumber, and processed foods

20	25	BONUS	NOTES
The largest state in the United States whose indigenous peoples have largely replaced their traditional ways with living and working in urban areas	The Mile High City, which was once a railroad and mining town and is now a center of transportation, finance, and of the steel and aerospace industries	A growing tourist industry that brings revenue to certain snow-covered mountainous areas	
The Great Basin state that derives much of its revenue from the gambling industry and nightclubs	The source that supports much of the West's economy because it owns or controls vast forests, parks, reservations, and space and power projects	The port on Prince William Sound, the southern terminus of the Trans-Alaska Pipeline, which carries petroleum from the Alaska North Slope	
This sparsely populated state specializes in cattle and sheep raising and wool production	The most important mineral in the West, half of which is mined in Arizona	The primary fruit that is grown and processed for export on Hawaii's tropical plantations, as is sugarcane	
Term for the climate type that enables California's valleys to grow oranges, olives, grapes, figs, almonds, and walnuts	The polluting condition caused by automobile exhaust fumes and factory smoke, especially in Los Angeles where car travel and manufacturing are the highest in the United States	The first and largest of the many U.S. national parks, located mostly in northwestern Wyoming	
A huge gorge, especially deep and colorful, in northern Arizona, which the Colorado River has cut through the Colorado Plateau	A string of islands extending from the Alaskan Peninsula and enclosing the Bering Sea	The West's largest lake, near the Mormon settlement, named for minerals deposited but not drained away	

UNIT 3

Canada

Canada
Topic: Physical Features

3

	LAND	WATER	CLIMATE AND SOIL	NATURAL FEATURES	PEOPLE AND PLACES
5	The vast, rocky, lake-dotted, unfarmable upland, surrounding Hudson Bay like a giant horseshoe	A chain of five freshwater lakes through which (except for Lake Michigan) runs the U.S.-Canada boundary	The treeless region in the far North where the winters are very cold and the short summers are cool	One of the world's best fishing grounds, located southeast of Newfoundland, that became seriously depleted in the 1980s	The native inhabitants of the Arctic tundra who support themselves by hunting and fishing, many of whom now live and work in fixed settlements
10	Beautiful mountains extending north from the United States, bordered on the east by plains and on the west by the Coast Mountains	An inland sea in northeastern Canada, forming a hollow center of the Canadian Shield	The large area of poor soil and rock that is extremely rich in minerals, including copper, gold, iron, lead, nickel, platinum, silver, uranium, and zinc	Canada's chief source of power that, where abundant and cheap, promoted industrial development	The predominantly French province where differences in education, culture, and civil policy have led to ideas of secession and independence
15	The bleak, cold region north of the Canadian provinces that includes the Arctic islands	Famous waterfalls on the river connecting Lakes Erie and Ontario that are a source of area power	The agricultural product raised west of the Wheat Belt in the drier short-grass region	The most basic, necessary natural resource, of which Canada has one third of the world's supply	An environmental problem caused by sulfur compounds from power plants, gas wells, and refineries

The province created in 1999 where most of Canada's Inuit live

A radioactive mineral found chiefly near Lake Athabasca in the Canadian Shield

Term for the climate type near Vancouver, where the abundant rainfall and mild year-round temperatures permit the growing of a variety of crops

An arm of the Atlantic Ocean between New Brunswick and Nova Scotia famous for its tides, which are among the highest in the world

Canada's easternmost province, consisting of an island and Labrador, which derives its income from its fishing, lumber, and mining industries

20

The type of industry that prompted the creation of small settlements in the bleak, rocky Canadian Shield and the Rockies

The most important catch of the fishing industry along the Pacific coast of British Columbia

The most densely populated and productive region, due to milder climate, a longer growing season, and proximity to power and transportation

A river system that drains Athabasca, Great Bear, and Great Slave lakes and that forms a water transportation route to the Arctic Ocean

The name for the Atlantic provinces of New Brunswick, Newfoundland, Nova Scotia, and Prince Edward Island

25

An agreement to eliminate tariffs and other barriers to trade among Canada, the United States, and Mexico

The Newfoundland site of a gigantic hydroelectric power project

Permanently frozen soil lying a few feet beneath the surface of the tundra

The river, rising in the Canadian Rockies and flowing through a deep valley to the Pacific at Vancouver, that is harnessed for electric power

Canada's northernmost point, 500 miles (800 km) from the North Pole

B O N U S

N O T E S

3 Canada

Topic: Occupations

FARMING AND FORESTRY	MINING	INDUSTRIES AND PRODUCTS	TRANSPORTATION	CITIES
5 The main food and export crop of the western plains, benefiting from the level land, modern methods and machines, and newer seed	The metal mined near Sudbury, Ontario, that the United States needs and that Canada leads the world in producing	A source of income in the Maritime Provinces, made possible by the area's sandy beaches, picturesque villages, and well-stocked fishing streams	The system of canals and locks, built jointly by the United States and Canada, that connects the Great Lakes with the Atlantic Ocean	The large industrial city, situated on an island in the St. Lawrence River, that is a center of shipping trade
10 The resource that covers half of Canada, yet is largely untapped because of distance from transportation and markets	A precious metal used in photography, coins, jewelry, and tableware, mined in Ontario, Québec, and Alberta	The best-known product of the "Golden Horseshoe," the industrial region surrounding the western shore of Lake Ontario	Three canals, two in the United States and one in Canada, that connect Lake Superior with Lake Huron and bypass the rapids in St. Mary's River	A huge industrial city and port on Lake Ontario, important as a financial and transportation center and for its printing and publishing industry
15 The name given to the provinces of Manitoba, Alberta, and Saskatchewan, where the fertile plains produce wheat, barley, oats, and livestock	High-grade ore found at the Québec-Labrador border, which may be important to the United States as a replacement for Minnesota's depleting supply	A metal needed for related industries, produced by mills in Hamilton and Sault Ste. Marie, Ontario, and in Sydney, Nova Scotia	The waterway used by oceangoing vessels to pass between Lakes Erie and Ontario, which are naturally joined by the Niagara River and Falls	A major city in thriving British Columbia, whose busy harbor is protected by a large island with the same name

The capital of Canada, located on the border of Québec and Ontario, many of whose people work in government or wood-related industries

The only walled city in North America, beautiful and historic, as well as a seaport and manufacturing center

An international port on Lake Superior that ships area wheat and manufactures transportation products

The key commercial, transportation, and distribution center for the agricultural and mineral products of the southern plains

The Nova Scotia city with an ice-free harbor that becomes an important winter port when the St. Lawrence River freezes

The transcontinental road, opened in 1962, that runs almost 5,000 miles (8,045 km) from St. John's, Newfoundland, to Victoria, British Columbia

A mineral used in the production of fertilizer, yielded by a fast-growing mining industry in Saskatchewan

The industry in which the majority of Canadians are engaged, working to meet the needs of others

One of the world's largest aluminum production centers in British Columbia, located near a huge hydroelectric project

A large island off the east coast of Nova Scotia and a part of that province, having rich coal and gypsum deposits

A metal Canada leads the world in producing that is used for galvanizing and making brass

Sand deposits that can be processed into petroleum, with the Athabasca River area having the world's largest deposits

Two chief products and leading exports of the softwood forests

The type of farming in Ontario and Québec that supplies a large home market

The Garden Province, so called because much farming is done on its fertile red soil

20

25

B O N U S

N O T E S

UNIT 4

Middle America and the Caribbean

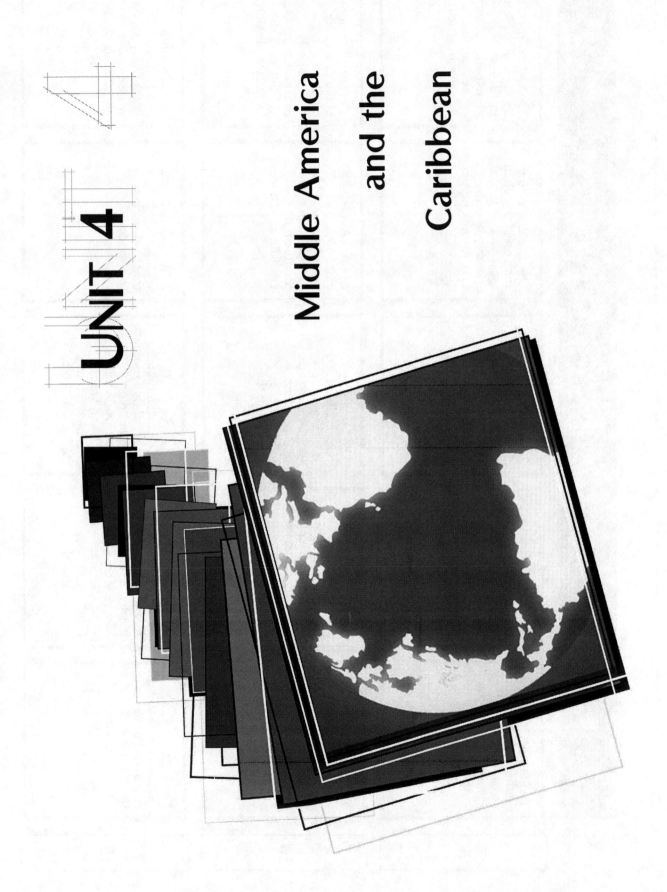

Middle America and the Caribbean

4

Topic: Physical Features

	LAND	WATER	CLIMATE AND SOIL	NATURAL RESOURCES	POTPOURRI
5	The largest island in the West Indies, most of which is low-lying and under cultivation	The body of water separated from the rest of the Atlantic Ocean by the West Indies	The group of island nations in this region that have similar plant life and the same tropical climate	The country having such great petroleum reserves that it could possibly become a major world supplier	The waterway built by the United States in 1914, an engineering marvel, connecting the Atlantic and Pacific oceans
10	The two mountain ranges that border the Central Highlands of Mexico on the east and on west	The river that forms 1,300 miles (2,092 km) of the border between the United States and Mexico	The feature responsible for the cool, comfortable temperature much of this region enjoys, although it lies within the low latitudes	Central America's chief source of power, provided by dams and power plants on its swift-flowing rivers	The island composed of the physically similar but economically different countries of Haiti and the Dominican Republic
15	The long Mexican peninsula extending southward from the U.S. state of California, which receives very little rain	An elongated arm of the Pacific Ocean that reaches into northwestern Mexico and almost separates Baja California from the mainland	The type of vegetation found in much of this region that is being destroyed and replaced with ranches, mining, and plantations	A fuel, often found with petroleum, that is abundant in Mexico	An independent commonwealth in the West Indies whose residents are citizens of the United States

The most densely populated section of Mexico, where redistribution is needed to remedy overcrowding

The nation that controls a group of islands near South America, including the islands of Curaçao and Bonaire

A person of Spanish or French ancestry born in the West Indies or Latin America, and the language of these people

An important export crop, taken from tropical trees, that is used to manufacture chewing gum

A leaf fiber, grown on the tropical lowlands of Yucatán, that is used to make rope and twine

A black, hard, heavy, durable wood from Mexico and the West Indies

Violent storms that often strike the West Indies during the rainy season and cause massive harm and great changes to the area's economy

Destructive natural forces that repeatedly destroy cities and economies in Central America

The agricultural use of the desertlike northern half of Mexico

The huge body of water on which are located Mexico's chief seaports, Veracruz and Tampico

The lake that makes up more than one third of the Panama Canal

The largest lake in Central America, surrounded by grazing, farming, and urban areas

20

The large, flat, dry Mexican peninsula whose abundant rains are absorbed by limestone

25

The group of hundreds of islands southeast of Florida that are composed of coral reefs and sand

B O N U S

A British colony of hundreds of coral islands lying east of North Carolina, warmed by the Gulf Stream

N O T E S

Middle America and the Caribbean

Topic: Occupations

4

	Agriculture	Mining	Manufacturing	Tourism	Government Influence
5	The most important food crop, occupying much of the farmland of Mexico and Central America	The island nation that produces its own oil and refines oil from the Middle East	Mexico's chief center of heavy industry, specializing in iron and steel and other metal processing	A major ingredient, in addition to good accommodations and beautiful beaches, that makes the West Indies a popular world resort	The French-speaking Caribbean country where most people have African ancestry
10	A major export crop grown on the coastal lowlands of Central America and in the West Indies	A world leader in the production of silver, which also has deposits of gold, copper, lead, and zinc	The refined product of the leading industry of the West Indies, which also yields molasses and rum	A new source of income for parts of this region having rain forests and natural wonders, generated by a growing awareness of the environment	Two islands, once British possessions, lying off the east coast of Venezuela, that have united and formed one independent nation
15	A plantation crop of Central America's volcanic highlands, which requires much hand labor	One of the world's major producers of bauxite, which also has supplies of gypsum and silica	Factories that receive parts from American firms and ship back assembled goods to the United States	The country, independent since 1981, whose huge spectacular coral reef and lush rain forests bring in tourist income	The unfortunate situation in many of these nations that has crippled economies, destroyed lives, and halted development

20

- The plantation crop grown in many of these countries that is a major export of Cuba, along with cigars
- A nonmetallic element used for explosives, medicines, and vulcanizing rubber, which is a major export of Mexico
- The Caribbean island that has the highest standard of living because it has received much industrial assistance from the United States
- Types of products made by natives largely for the tourist trade
- The economic area in which Latin Americans feel foreign investment is more urgently needed than in oil and mining

25

- An important textile crop used by natives and sold for export
- The mineral used in glassmaking and graphite, of which Mexico is the world's largest producer
- A valuable tropical hardwood, used for furniture, that is found in Mexico and the West Indies
- The area's attractions for tourists and archaeologists to which these names refer: Aztec, Mayan, and Toltec
- The government program that has made much of Baja California's arid land agriculturally productive

BONUS

- One of the Western Hemisphere's tiniest nations, often called the Isle of Spice for its nutmegs, which also yield mace
- A major world supplier of natural asphalt, which also has large petroleum reserves
- A leading export crop of Cuba and other West Indian islands that is used for cigars and cigarettes
- A group of three islands and many islets in the West Indies that make up a U.S. territory famous for tourism
- The country sometimes resented by Latin Americans because of its strong political, economic, and cultural influence

NOTES

Middle America and the Caribbean

4

Topic: People

	PEOPLES	CITIES	PROBLEMS	SOLUTIONS	CHANGES
5	A person of mixed black and white racial backgrounds	The largest city in the Western Hemisphere, and the center of its nation's economic, industrial, political, and cultural life	The cause of continuous governmental need to provide more housing, additional jobs, and larger schools	The organization of Latin American countries and the United States that works for common defense and cooperation within the bounds of the United Nations	The country that took over the administration of the Panama Canal on January 1, 2000
10	A person having mixed European and Native American ancestry, as most Mexicans do	The largest city and capital of Cuba, and the center of its industry and export trade	A major cause of poor crop yields from small farms due to lack of money and education	The solution thousands of Latin Americans have sought in order to escape unemployment, overcrowding, and government brutality	The large Caribbean nation, supported by the U.S.S.R. until 1991, where harsh Communist control has created severe shortages of food, clothing, and fuel
15	The race of most West Indians, whose ancestors were imported by Spaniards to work on plantations	The capital of Puerto Rico, which is a major manufacturing center in the West Indies, having many U.S. plants	One of the most air-polluted cities in the world, where the mountains prevent winds from blowing away factory smoke and vehicle exhaust	The major road system to which the International American Highway belongs, Central America's only road link among its nations	The country whose population is concentrated around its two large lakes where the 1997 election was the first peaceful transfer of power in the nation's history

The smallest, most densely populated Central American country whose agricultural and industrial economy was slowed by years of civil war, drought, and natural disasters

The country whose high standard of living includes generous social services, no armed forces, a high literacy rate, and an extensive national parks system

Institutions that are required but that often are too few or too distant or out of reach for those who must help support their families

An international banking city and transfer point for trade and transportation

The predominant language of this region and a major cultural influence

20

The least developed and poorest of Central American countries, which is attempting to rebuild after the devastation caused by Hurricane Mitch in 1998

The government program through which Mexico has redistributed half of the farmland of haciendas to individuals or to community groups

Housing that results when too many people move from farms to find homes and jobs in crowded urban centers

The densely populated country of which San José is the capital, whose economy has prospered because of rich volcanic soil and adequate hydroelectric power

Farming in which only simple food crops are raised for family use

25

The island that seceded from the Netherlands Antilles in 1986

An agreement to eliminate tariffs and other barriers to trade among Canada, the United States, and Mexico

The problem of workers on sugar plantations, where the harvest lasts six months and the rest of the year is the "dead season"

The city, other than Mexico City, that has the largest urban population in Central America and that is the capital of its country

The severely poor, coffee-growing Central American nation half-populated by Native Americans who still cling to their traditional ways, culture, and language

B O N U S

N O T E S

Geography Challenge!   35

UNIT 5

South America

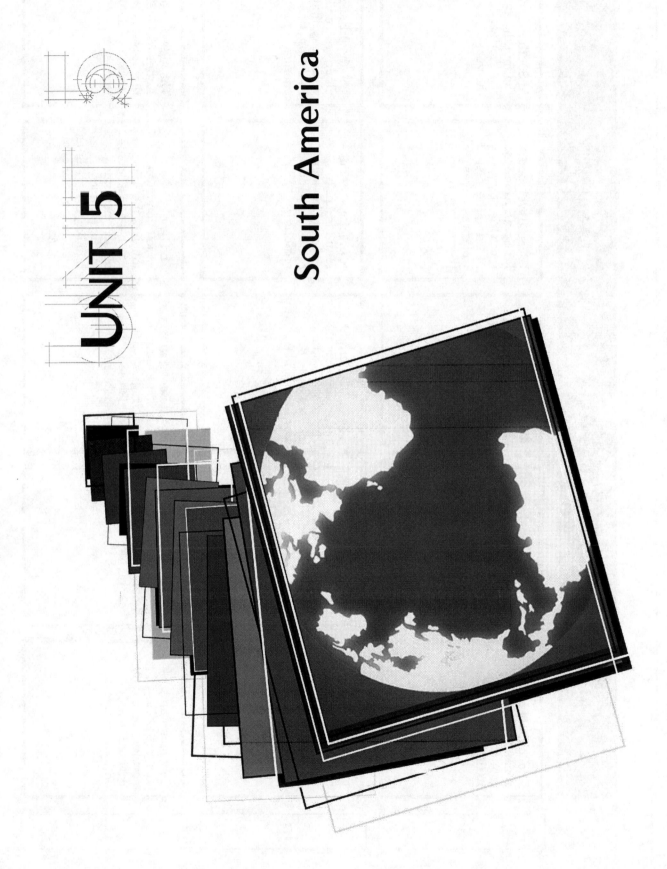

South America
Topic: Physical Features

	LAND	WATER	CLIMATE	NATURAL RESOURCES	POTPOURRI
5	The longest and second highest mountain chain in the world	The river, 3,900 miles (6,275 km) long, that forms the largest drainage basin in the world	The 2,000-mile-long (3,218 km) desert on the coastline of Peru and northern Chile	Power developed for South American industries by utilizing the vast water resources there	In South America, the month having the shortest day of the year
10	Mountains in eastern Brazil along the Atlantic Ocean	The estuary of the Paraná and Uruguay rivers, extending 225 miles (362 km) inland, on which Buenos Aires is situated	The ocean current responsible for the dry coastal lands of Peru and northern Chile	A precious metal found in large deposits in each of the highland regions	The disturbances, along with active volcanoes, that cause physical change in the Andes Mountains
15	Low mountains along Brazil's northern boundary, covered mainly with tropical forests	The large shallow lake in Venezuela that is the site of vast oil deposits	The treeless, grassy plains in Argentina and Uruguay that have rich soil and a temperate climate	The type of vegetation covering much of the Amazon Basin—a rich source of food, perfume, medicine, and building materials, as well as the habitat of animals	The side of the Andes Mountains that receives rain from the Atlantic trade winds

5

20	25	BONUS	NOTES

20

- The broad, high plateau between parallel ranges of the Andes in Bolivia and Peru
- The highest body of water in the world on which steamships operate, located on the border of Bolivia and Peru
- The flat, rocky tableland in southern Argentina, having a cool, dry climate
- A group of islands, 650 miles (1,046 km) west of Ecuador, on which live huge turtles and various strange animals
- The element that is in short supply in the upper elevations of the Andes Mountains, making breathing difficult

25

- The frigid island shared by Argentina and Chile at the tip of South America that is beset by dangerous winds and seas
- An important river in Venezuela that empties into the Atlantic Ocean through a huge delta
- The level grassland region in the tropics
- The condition of the soil in the rain forest
- The partly forested, sparsely populated scrublands in Paraguay and Argentina

BONUS

- The highest peak in the Western Hemisphere, at 22,834 feet (6,850 m) above sea level
- The channel of water that separates Tierra del Fuego from the South American mainland
- A major feature that moderates the South American tropical climate
- The South American beast of burden that is also a source of fine wool
- The highest waterfall in the world (3,212 feet or 979 m), located in Venezuela

NOTES

South America

Topic: Occupations

5

	AGRICULTURE	FISHING AND FORESTRY	MINING	MANUFACTURING	MORE INCOME
5	The major agricultural industry in South America, found mostly in Patagonia	One of the world's largest fishing nations, dealing in tuna, and in anchovies for fishmeal	Once South America's richest country and the world's largest exporter of petroleum, it suffered financially when oil prices dropped in 1986.	South America's major nonfood products that are manufactured from native-grown raw materials	The highly profitable cash crop of poor Andean farmers that is the raw ingredient of cocaine
10	A crop requiring much hand labor, grown chiefly in the São Paulo area of Brazil	A tree in coastal lowlands that yields seeds used in the manufacture of cocoa and chocolate	The country that is a world leader in copper production and exporting	A leading grain of the pampas that is made into flour	The leading textile product exported from the ranches of Argentina and Uruguay
15	A plantation crop of the rain forests of Ecuador, Brazil, and Colombia	An excellent furniture wood important to the timber industry in Colombia	The country that is one of the world's largest producers of tin	The ore from which aluminum is produced, found mostly in Suriname and Guyana	Flat grasslands north of the Orinoco River where livestock raising is important

20	Animals raised on huge ranches for their hides and processed as meat for export	A tree growing wild in the Amazon Valley that caused a boom in growth and wealth in the late 1800s and is still grown on plantations there	An abundant ore, found largely in Brazil and Venezuela, that needs coking coal for development

The country that is a major vacationland because of its soccer, performing arts, and miles of beaches

An important export produced from a crop grown on Brazil's and Guyana's plantations

25

A crop grown in Brazil and Peru for export and for the textile industry

A very tall nut tree that grows in the Amazon rain forest

The country that is a leading producer of silver, lead, zinc, and copper, where petroleum and natural gas have added to the economy

The country that has the world's only deposits of natural sodium nitrate, used to make fertilizer and explosives

The organization to which Venezuela and Ecuador, as oil-producing nations, belong

B O N U S

The name given to farms where only simple food crops are grown for the family

The characteristic of fine tropical wood that makes it difficult to ship out of the rain forests

One of the world's most expensive gems, of which Colombia mines half of the world's supply

A major advantage that South America has in the manufacturing industry

The occupation that before 1950 involved more than half and now involves about a quarter of South America's labor force

N O T E S

5 South America
Topic: People

	PEOPLES	CITIES	PROBLEMS	CHANGE	CLOSING THE GAP
5	A person of mixed Native American and Spanish descent	The largest city in South America and one of the largest and fastest-growing cities in the world, with its industries attracting native and foreign immigration	The huge resource that is being rapidly depleted as the interior is opened up for settlement, development, and ranches	The only colonial territory among 12 independent countries in South America	The Latin American countries and the United States, working within the framework of the United Nations for defense, cooperation, and peaceful settlement of controversies
10	The race to which most people in the southern part of the continent belong	Argentina's capital and one of the world's largest ports, located on the Río de la Plata 170 miles (274 km) from the Atlantic Ocean	A source of wealth as well as violence and corruption in Bolivia, Colombia, and Peru	A small country that gained its independence from Great Britain in 1966	A system of highways extending from the U.S.-Mexico border to Chile and linking the capitals of 17 Latin American countries
15	The largest racial group in eastern Brazil and coastal Venezuela	The capital and chief city of Venezuela, which has grown from an old colonial city to a metropolis because of the country's booming oil industry	The living situation created around cities by the influx of the uneducated and unskilled seeking a better life	The country once called Dutch Guiana that became independent in 1975	A major problem being solved by the modernization of transportation and communication

20	25	BONUS	NOTES
The answer to long-distance travel and travel over South America's difficult terrain	A country of contrasts, having the most successful economy in South America due to tourism, manufacturing, industries, wineries, and agriculture	Latin America's most modern country, largely European, that in 1989 restored democracy and organized a financially stable government after decades of decline	
A major need of South America, without which the building of a self-supporting economy is handicapped	The answer for many South Americans in their quest for social justice, prosperity, and economic independence	Once poor despite its rich land and pleasant climate, this small country is now thriving due to smuggling and the sale of electricity	
A major cause of poor communication, besides the inability to own telephones, radios, and television sets	The primary need that South Americans must meet in order to keep up with the growing population	The two landlocked countries in South America that lost much of their valuable territory in wars in the 1800s	
The "Athens" of South America, and Colombia's center of government, arts, and education	One of the world's most modern and beautiful cities, and the center of Brazilian life	The modern planned city that was built to draw population away from the crowded coast and that has served as its country's center of government since 1960	
The race of much of the population of inland and mountain areas	The country that combines the cultures of the Portuguese, indigenous peoples, and people of African ancestry	The official language of most of the South American countries and of almost half South America's people	

20 **25** **B O N U S** **N O T E S**

UNIT 6

Europe

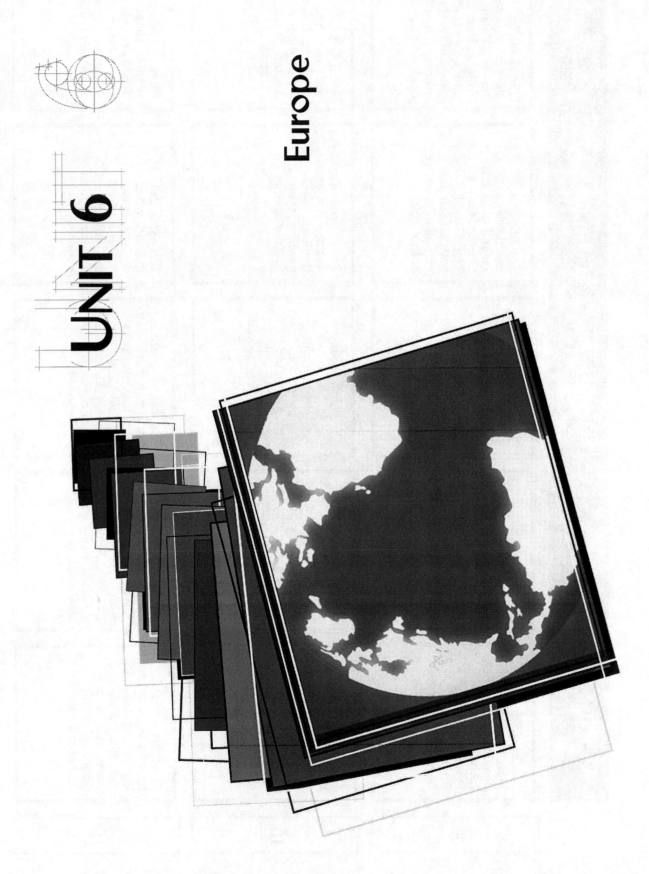

Europe
Topic: Physical Features

6

LAND	WATER	CLIMATE AND SOIL	NATURAL RESOURCES	POTPOURRI
5 Southern Europe's high, rugged, beautiful mountains, including the Pyrenees, Alps, Apennines, and Carpathians	Europe's major commercial waterway, used also for power, agriculture, home use, and recreation	The type of climate shared by the Iberian, Italian, and Balkan peninsulas, where the summers are hot and dry and the winters are cool and wet	The important coalfields that Poland gained from Germany after World War II, which support the iron and steel industry of the region	The name for the Nordic countries of Norway, Sweden, Finland, Denmark, and Iceland, which have built a sound economy despite natural difficulties
10 The high, jagged mountain range forming the border between France and Spain and containing the small country of Andorra	The long water route joining western and eastern Europe, navigable for its entire length from western Germany to the Black Sea	The warm ocean current that flows northeast from the Caribbean Sea, moderating the temperatures of western Europe and keeping the ports ice free	Power used in many European countries that lack coal, enabling their cities to have cleaner air	A low-lying area reclaimed from the sea by enclosing it with dikes and draining it with pumps
15 The long, narrow country bounded on the east by rugged mountains and on the west by a long coastline deeply indented by fjords	The sea in northern Europe that serves the Scandinavian seaports of Stockholm and Helsinki and the Polish port of Gdansk	The low-lying, almost flat region, stretching from the Atlantic to the Urals, that has some of the most productive farmland in the world	The capital of Iceland, which derives its steam power for heat and light from underground hot springs	An early Italian seaport whose famous canals and art treasures are threatened by air and water pollution

The sea that the Delta Project, completed in 1986, holds back from overflowing southwestern Netherlands

The world's largest island, which lies near North America and is basically a huge ice sheet, and which has gained home rule from Denmark

The animal the Saami of northern Scandinavia herd and depend on for food, clothing, and transportation

Brown coal, less valuable than black, used to generate industrial power in eastern Germany and the Czech Republic

The type of power that generates most of France's electricity

The first stage in the formation of coal—spongy, decaying plant matter—that is used as a fuel source in Ireland and Finland

The hot, dry plain of central Spain and Portugal where wheat is grown and sheep are raised

The large island at Italy's southern tip, where citrus fruits and olives are raised in the fertile volcanic soil from Mount Etna

Dust and silt blown from glaciated areas and deposited in thick beds, forming very fertile soil

The long arm of the Mediterranean Sea bordered by Italy on the west and the Balkan Peninsula on the east

The river that forms a valley across northern Italy, the most industrial and agricultural region of the country

The Land of Thousands of Lakes that conducts much of its foreign trade with its neighbor on the east—Russia

A mountainous country composed of several peninsulas and many beautiful islands in the Aegean Sea

The rugged mountains, forming the "spine" of Italy, whose northeastern slopes surround the tiny country of San Marino

A northern country composed of the Jutland peninsula and an archipelago to the east, connected by ferries and bridges

20

25

B O N U S

N O T E S

Europe
Topic: Occupations

AGRICULTURE	FORESTRY AND FISHING	MINING	MANUFACTURING	MORE INCOME
				A country famous for its manufacture of valuable precision instruments and timepieces, using little raw material and great technical skill
			Famous fine-quality textiles produced in the United Kingdom from native and imported raw materials	
5 — Europe's chief grain crop, grown almost everywhere because its many strains are adapted to a variety of soils and climates	The major resource and export of Finland and Sweden, where hydroelectric power is used to process it into paper, pulp, and furniture	The mineral deposits around which industrial areas have developed in Europe		
				A leading agricultural nation known for high-grade dairy products and ham, which come from its efficient farms run by cooperatives
			The classification of products such as jewelry, perfumes, and women's high-fashion clothing, made in Paris and other parts of France	
10 — The main product made from the many varieties of grapes grown in France and the Mediterranean countries	A Nordic nation whose ice-free harbors enable it to earn much income from fishing and shipping	One of the world's most important coalfields and industrial areas, located in western Germany		
				One of Europe's most famous resort regions, lying between the Alps of southeastern France and the Mediterranean Sea
			The tiny, densely populated, strategically situated country in the Mediterranean Sea that specializes in shipbuilding and repair	
15 — A source of sugar and fodder, grown chiefly in France, Germany, and eastern Europe	The season during which trees are cut in northern forests	The leading mineral of France that is used with coal for the heavy industries of north-central Europe		

The republic, located on an island west of England, whose agricultural economy has largely been replaced by an economy based on computers and technology

The type of agriculture carried on by the Dutch on the damp, rich polders

The world-famous cash crop of the reclaimed lands of the Netherlands that also serves as a tourist attraction

The plant from which Belgium and Northern Ireland make fine-quality expensive linens

A country known for its fine automobiles, racing cars, and other highly technical products such as cameras and computers

In Germany's chemical industry, a mineral used as a base for plastics, dyestuffs, drugs, and fertilizers

The body of water in northern Europe where offshore drilling for oil and natural gas has been going on for some time

The Low Country whose coalfields have promoted the steel and textile industries and whose city of Antwerp is a busy port

A nation whose high standard of living is aided by an urban industrialized economy based on extensive forests, rich iron ore deposits, and abundant water power

A sparsely populated island nation where fishing, fish processing, and sheep raising are the main industries

The poor, hilly country known for its port wine, cork, and sardines

Endangered sea animals, once an important source of food, oil, and by-products, the hunting of which has sharply decreased

A staple food of Poland, which grows a large part of the world's supply on its poorer, cooler plains

The grain that ranks second as a source of bread flour, much of which is raised in northern and eastern Europe

A Mediterranean crop, the finest quality coming from Spain, grown as a food and a source of cooking oil

20

25

BONUS

NOTES

Europe
Topic: People

	PEOPLES	CITIES	TRANSPORTATION AND TRADE	GOVERNMENTS	POTPOURRI
5	The country where Protestants wishing to remain aligned with the United Kingdom have clashed with Catholics who wish to join the Republic of Ireland	The largest European city, a chief port on the Thames River, and the center of British government, manufacturing, and finance	An organization of European nations, formed to unite the economic resources of member nations into a single economy, and strengthen cooperation among them	The name applied to the political unit that includes the countries of England, Scotland, Wales, and Northern Ireland	The world's first industrialized nation, whose ideal location for world trade and protection from destructive warfare has enabled it to prosper
10	The smallest independent state in the world and the center of the Roman Catholic Church, located within the city of Rome	The world center of the arts and education, which lures tourists with museums, gardens, shops, and historic buildings	The new tunnel under the English Channel that connects England and France, linking Great Britain with the continent	The country reunited in 1990 after being divided for 45 years into two separate republics: the free western section and the Communist eastern section	The gigantic problem that must be addressed by cooperating European nations in order to protect their drinking water, clean air, and forests
15	The tiny country located on the French Mediterranean coast that is famed as a resort and for its huge gambling casinos	The capital of Italy, an ancient as well as modern city on the Tiber River	The seaport at the mouth of the Rhine River that handles more cargo than any ocean port in the world	The country on the Balkan Peninsula that split into several republics in 1995: Macedonia, Slovenia, Croatia, Yugoslavia, and Bosnia and Herzegovina	Organizations of farmers who share costs, machines, and profits for more efficient production

The region of Scandinavia above the Arctic Circle where the nomadic Saami live

The capital and leading manufacturing city of Greece, best known for its historical ruins, which attract many tourists

The means by which northwestern and southern Europe carry on 75 percent of international shipping

Germany's largest city and once again its capital, reunified after the fall of the Berlin Wall

The world position of Switzerland and Austria, in which they take no part in treaties and little in world affairs

The name for traveling farm laborers, many from the Mediterranean countries, who help plant and harvest crops in western Europe

The colorful Danish capital, through which most ships enter and leave the Baltic Sea

Internal avenues of transportation, especially in France, Germany, and the Netherlands, that were built to carry goods economically

The separate states formed when Czechoslovakia split into two republics in 1993

The eastern European country that was reshaped and rebuilt after World War II and is now struggling to create a free market economy after many years of Communist control

The term pertaining to a large group of people sharing the same culture, language, religion, or ancestry

Austria's capital on the Danube River, once a center of music, culture, and learning, and now a rebuilt industrial city

A wide, deep waterway that joins the rivers of eastern and western Germany and gives Berlin access to the Atlantic Ocean

The tiny British Crown Colony that controls the narrow western entrance to the Mediterranean Sea

The poorest, least developed Balkan country, isolated from its neighbors by the Dinaric Alps

20

25

B O N U S

N O T E S

UNIT 7

Northern Eurasia

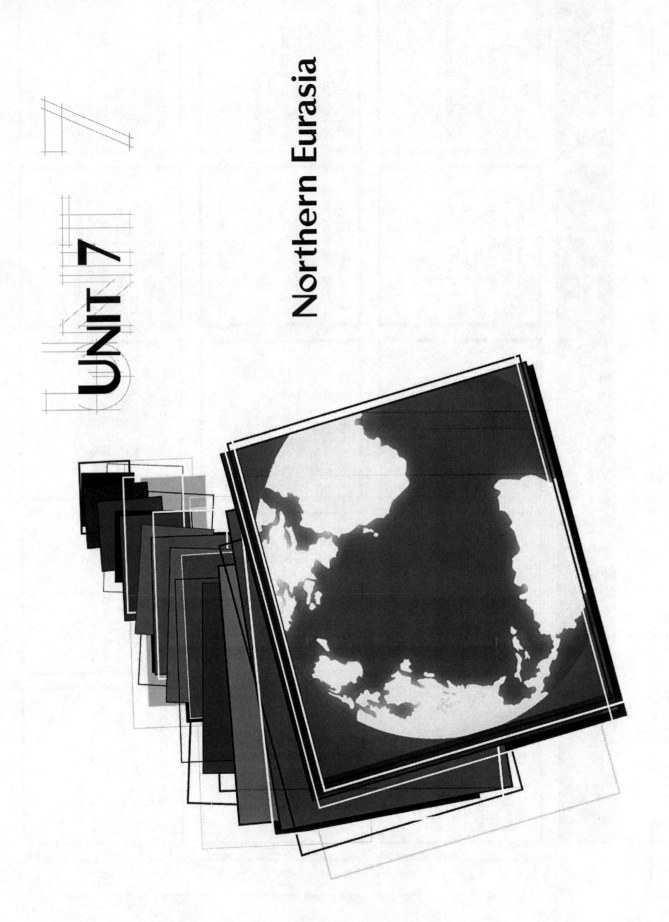

7 Northern Eurasia
Topic: Physical Features

	LAND	WATER	CLIMATE AND SOIL	CITIES	INTERESTING FACTS
5	The name of the low mountain range and the river that, with the Caspian Sea, separate the continents of Europe and Asia	The wide, slow-moving river that rises in the Valdai Hills and empties into the Caspian Sea, connected by canals to the Don River and the Baltic Sea	The treeless, barren land along the Arctic coast having a frigid, harsh climate	Russia's largest city, serving as its capital and cultural center, and producing vehicles and textiles	The federation formed from 11 of the 15 former republics of the Union of Soviet Socialist Republics (U.S.S.R.)
10	The high, rugged mountains stretching between the Black and Caspian seas that form the southern boundary between Europe and Asia	The world's largest lake, salty and below sea level, located in the dry sandy region south of the Urals	Fertile grasslands extending from Ukraine to Siberia that are intensively farmed, especially with grains	An ancient caravan city in Uzbekistan, now a center of the modern textile industry	The largest of the 11 republics in the C.I.S., containing two thirds of the land area and one half of the population
15	An unofficial region encompassing most of Asian Russia, known for its long, severe winters and rich mineral reserves	The southwestern body of water, bordered also by Turkey, Bulgaria, and Romania, into which the Dnieper River flows	The climatic region east of the Caspian Sea and along the south's mountainous border, where rivers are used to irrigate farmlands	A Pacific port in southeast Siberia, kept open by icebreakers, that is a base for fishing fleets and is the eastern railroad terminal	The three small republics on the Baltic Sea that declared their independence from the Soviet Union in 1991

A southwestern lake, now shrunk to one third its former size through irrigation, around which much plant and animal life has been devastated

The year that the U.S.S.R. and Soviet Communism ceased to exist

The system of government that regulated the Soviet people until 1991

An important seaport and railroad and industrial center located on the Don River, near where the river empties into the Sea of Azov

A rapidly growing city in western Siberia that is a center of transportation and industry

The new country on the Black Sea, of which Tbilisi is the capital, that is protected by the Caucasus Mountains

The great coniferous forest belt from Norway to the eastern coast of Russia, where the climate is severe and the soil is poor

The warm summer/cold winter type of climate in much of this region, becoming more extreme inland

Rich, deep, black soil of the steppe, ideal for growing wheat and other grains

The long river that flows from the southern mountains northward into the Kara Sea and that, with its tributary the Irtysh, drains the Siberian plain

The narrow body of water that joins the Bering Sea with the Arctic Ocean, and the point where Russia and the United States are only 50 miles apart

The earth's oldest and deepest lake, which contains one fifth of the planet's freshwater and which remains largely unpolluted through the efforts of many

The type of land found in the western half of this region, broken only by the low Ural Mountains

A long volcanic peninsula in the northeast of Russia that, with the Kuril Islands, encloses the Sea of Okhotsk

The highest point in this region, rising in the Pamirs in Tajikistan

20

25

B O N U S

N O T E S

7 Northern Eurasia

Topic: Occupations

AGRICULTURE	FORESTS, FISH, AND FURS	MINING	INDUSTRIES	TRANSPORTATION AND TRADE
5 The leading grain of this region, yet not enough of it is produced for the region's needs and more must be imported	The abundant resource of the northern lands that is processed into paper, especially in the European section of this territory	The mineral found in huge deposits in the Donets Basin in Ukraine that has contributed to the area's industrial growth	The enormous industry based on the resources of the Donets Basin and Krivoy Rog (Kryvyy Rih) in Ukraine	The region's longest European river, dammed to facilitate travel, on which much internal trade depends
10 The plant yielding coarse fibers used for blankets and padding that is grown on the irrigated lands near the Caspian Sea	The aspect of the lumbering industry that makes it too expensive and too difficult to develop fully in the north Asian forests	High-grade ore deposits in the Ural Mountains, which are linked industrially with the Kuznetsk Basin	The underlying reason for expensive methods of construction needed to develop the abundant resources in Siberia	The chief means of transportation, which made possible the growth of manufacturing centers in many parts of the region
15 The agricultural use of the drier eastern steppe as well as the southern mountain lands	Two seas to Russia's northwest, arms of the Arctic Ocean, that are fished for cod, haddock, herring, and salmon	The area of rich coal deposits in western Siberia	A principal type of power, still largely undeveloped, that supplies the industries of Asian Russia	The "new" name for Leningrad, a major seaport built on the Gulf of Finland as a link with the West

20	25	BONUS	NOTES
The first rail route across Russia, connecting Moscow with Vladivostok, that is now part of a large network	The world's largest city north of the Arctic Circle that remains an ice-free port due to the North Atlantic Drift	The type of ship that cuts through frozen seas, enabling icebound ports to remain open for more of the year	
One of the oldest industries in the Moscow area and in Turkmenistan that uses native-grown materials	Manufactured products that continue to be in short supply during the region's transition from Communist mismanagement to an open-market economy	The republic on the Black Sea that ranks second to Russia as a great industrial and agricultural leader	
The fuel mineral found in Baku and Ukraine, but now chiefly taken from the Volga-Urals fields	The fuel source obtained in central Asia and in western Siberia at the Arctic Circle and piped to populated areas	An essential alloy metal used in steelmaking that is produced in large supply in Ukraine and Georgia	
The type of forest in the C.I.S. in Europe where the winter is milder, the rainfall heavier, and the soil richer than in the far north	The animal raised on the tundra and in central Siberia for food, clothing, and transportation	The expensive delicacy made from fish eggs which has led to the severe depletion of sturgeon caught in the Caspian Sea	
A basic food of most of the people, which grows in the poorer soils of the cool, damp regions	A versatile plant that yields both sugar and fodder	A cash crop grown in the steppe's black soil for cooking oil	

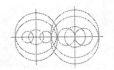

UNIT 8

Asia

Asia

Topic: Physical Features

8

	LAND	WATER	ISLANDS	CLIMATE AND SOIL	NATURAL RESOURCES
5	The system of high, rugged mountains along India's northern border, separating the Indian subcontinent from the rest of Asia	The northern part of the Indian Ocean, bordered by India, Bangladesh, and Myanmar, which serves the seaports of Calcutta, Madras, and Rangoon	A republic once controlled by the Dutch, consisting of many thousands of islands, including Sumatra, Java, Celebes, and part of Borneo and New Guinea	Seasonal winds that bring torrential rains in the hot summer and cool, dry air in the winter	The ore found in abundance in India that forms the core of major industries at Jamshedpur and in the northeast
10	The narrow peninsula in southeast Asia that includes parts of Myanmar, Thailand, and Malaysia	China's longest river, which is used for transportation and ocean commerce and which has millions of people living along its banks	The island nation off the southern tip of India, once called Ceylon, that raises tea and rubber on plantations	The term for the soil of the vast fertile northern plain of India, which includes the valleys of the Brahmaputra and Ganges rivers	The fuel mineral of which ample reserves have recently been found in Indonesia, China, and Malaysia
15	The dry region that dominates the Mongolian Republic, the outer margin of which is used by nomadic herdsmen	The greatest waterway in India, whose waters are used mainly to irrigate its fertile and densely populated valley	The developing Pacific republic of over 7,000 islands, the largest being Luzon and Mindanao	The dark-colored ocean current that flows northeastward from the Philippines, warming the climate of Taiwan and Japan	The growing concern in much of this area due to heavy coal use and unregulated waste water disposal

A major source of power in China, found in Manchuria and used in the iron and steel industry there

The fuel reserves recently discovered in Brunei and Indonesia, where large processing plants have been built

An alloy metal found in China, Korea, and Thailand that is used in steel and in electric bulb and tube filaments

A tropical cyclone occurring along the Pacific coasts of southeast Asia, accompanied by forceful winds and torrential rains

The natural occurrence that prevents the buildings in crowded Japanese cities from being built more than several stories high

Physical features of Indonesia and Japan that make local soil very fertile, thus attracting farmers to live there despite the danger

The largest and most populated of Japan's islands, suitable for both agricultural and industrial development

The most heavily populated and industrialized island of Indonesia, and the home of Jakarta, the country's capital and largest city

The second largest island in the world, whose western half is part of Indonesia

A great river flowing east, sometimes called China's Sorrow because its yellow waters cause ruinous floods in the North China Plain

The river that winds through Pakistan and into the Arabian Sea, used extensively to irrigate that country's deep fertile plains

The river that flows southward through Laos and Cambodia (Kampuchea) and empties into the China Sea near Ho Chi Minh City

The vast area, bounded by the Himalayas on the south and the Kunlun Shan on the north, that is too high, cold, and dry to support vegetation

A high, rugged area north of Pakistan from which several mountain ranges radiate, including the Tian Shan, the Himalayas, and the Kunlun Shan

The desert that lies in China's northwest Sinkiang region, surrounded by mountains that keep out rain-bearing winds

20

25

B O N U S

N O T E S

Asia

Topic: Occupations

8

	AGRICULTURE—CROPS	AGRICULTURE—METHODS	FISHING AND FORESTRY	MANUFACTURING	MORE INCOME
5	The area's chief crop and the basic food of the people, grown in the long, hot, wet summers	A flooded field, enclosed by dikes, in which wet rice is raised	The leading source of animal protein in the Japanese diet, as well as an important export product	A leading export of Japan that is derived from the cocoons of caterpillars raised on mulberry leaves	An occupation of millions who live along rivers and warm shallow seas, providing food for the natives as well as products for export
10	A plant grown on hot, rainy plantations in India and Sri Lanka for export, and on hillside farms in China and Japan for home consumption	Large estates set up in southeast Asia by Europeans to grow crops such as rubber and tea	A valuable tropical hardwood used for shipbuilding and fine furniture, grown in Indonesia, Myanmar, and Bangladesh	The resource that has enabled countries such as Japan, Singapore, Taiwan, and Hong Kong to become industrial giants	The country that is the world's largest producer of tin
15	The cash crop of the Deccan plateau on which India's major industry is based	The method used on steep hillsides for growing rice, utilizing all available land and water	The source of sea products that supplements the traditional catch in coastal and deep sea waters	The great industry the Japanese have developed from imported raw materials in order to provide for their ship, automobile, and machine production	The business of producing consumer goods at home, such as fine fabrics, carpets, brassware, and leather goods

20

A product used commercially worldwide that is derived from trees on hot, humid plantations in Malaysia and Indonesia

The agricultural product that overcrowded Asian countries cannot afford to raise on their limited farmland

The lumber industry process that is neglected in some tropical Asian countries, resulting in erosion and a dwindling wood supply

The area in northeastern China that is highly industrialized due to large deposits of oil, coal, and iron

The name for skillfully made artistic items that are produced for sale to tourists and markets throughout the world

25

A plant grown chiefly in the muddy Ganges Delta in India and Bangladesh from which twine and burlap are made

The animals many Chinese raise for meat because they feed on scraps rather than on grazing lands

The animal fished for in southeast Asian waters whose skeleton is used as a cleaning tool

The chemical product manufactured on the Indian subcontinent for the improvement of agricultural production there

Thailand's capital city, whose many shrines bring in tourist income

BONUS

A food crop and a source of oil and protein grown along with wheat on the large fertile plains of northeast China

The ancient farming method used in southeast Asia in which wooded areas are cleared by burning, then fertilized with the ashes

The special sea animal that is bred by the Japanese in tropical waters to produce cultured pearls

High-quality products made in Japan using few raw materials and much skill

A product of the coconut palm raised on Philippine plantations that yields an oil used to make soap, candles, and margarine

NOTES

Asia
Topic: People

8

	PEOPLE AND PROBLEMS	TRANSPORTATION AND TRADE	CITIES	GOVERNMENTS	POTPOURRI
5	The most populous country in the world, whose people, agriculture, and industry are crowded into the eastern section	The lifeblood of the economy of Japan, a country that needs to market goods in order to buy needed food and raw materials	The Westernized but still traditional capital and largest city in Japan, and its financial, commercial, and industrial center	The tiny prosperous coastal area in southeast China, one of the most densely populated areas on the earth, that reverted back to China in 1997 after 99 years of British control	The highest peak in the world at 29,028 feet (8,848 m) above sea level, located in the Himalayas
10	The resource near which the most densely populated areas are located because transportation and farming are convenient there	A tiny island republic that is a center for international banking and a major world port, which enjoys a high standard of living	China's largest city and a world trading and banking center, located near the mouth of the Yangtze River (Chang Jiang)	The island, also called the Republic of China, that is one of the world's strongest economies, operating apart from the Communist China mainland	Asia's major strength and hope for the future
15	The factor that has put an unbearable strain on the food supply, housing, schools, jobs, and medical help in most of these Asian countries	The organization for independence from foreign powers and for cooperation in strengthening members' capitalist economies	A large modern port in northeastern India, and a center of industry, engineering, and jute production	Three small countries, Communist as a result of a long war in the 1960s, whose people have been forced out by harsh rule, war, famine, or disease	The source of energy many Asian countries are turning to, because their need is great and their mineral fuel reserves are inadequate

The tropical giant grass with a hollow woody stem, used for homes, baskets, furniture, and fishing poles

The country whose two parts lie on the Malay Peninsula and on Borneo, and whose economy is thriving

The sacred river of the Hindus, to which thousands come to bathe and be purified

A hilly peninsula whose two nations—north and south—fought a costly war in the 1950s but are now seeking to improve relations

The incentive offered by Southeast Asian governments to encourage industrial growth, which some foreign investors have used to unfair advantage

The governmental system that controls resources, production, and welfare, whose regimes were greatly reduced by the political changes between 1989 and 1991

India's largest city, whose location on the west coast makes it an important seaport and commercial center

The capital and cultural, social, and commercial city of the Philippines, located on a superb harbor on the island of Luzon

China's beautiful capital city which has experienced rapid industrial growth

The manufactured product that is an export-import issue between the United States and Japan, its two top producers

One of the world's most famous mountain passes, linking northern Pakistan with Afghanistan

A wooden sailing vessel used for transporting goods on rivers and seas, and also used as a home for many along the Yangtze River (Chang Jiang)

The destination of millions of rural Asians who seek to escape poverty and strife

One of the most populated cities in the world, whose officials must solve severe pollution and traffic problems resulting from its rapid growth

A poor, struggling country that is beset by the flooding of its three major rivers and by monsoons and tidal waves

20

25

B O N U S

N O T E S

UNIT 9

Southwest Asia and North Africa

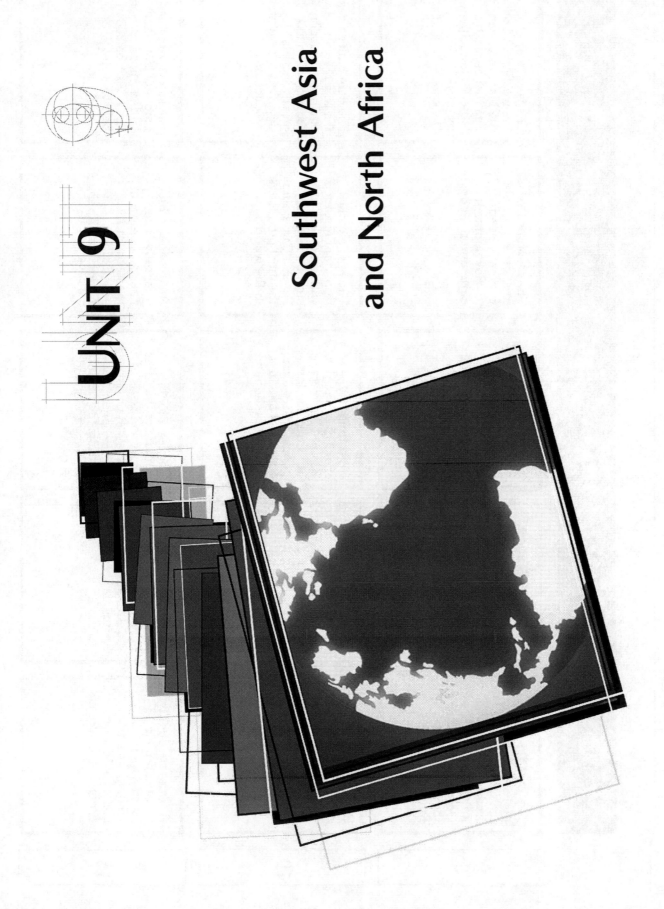

Southwest Asia and North Africa

Topic: Physical Features

9

	LAND	WATER	CLIMATE AND SOIL	NATURAL RESOURCES	POTPOURRI
5	The vast desert plateau that includes Saudi Arabia, Bahrain, Kuwait, Oman, Qatar, Yemen, and the United Arab Emirates	The world's longest river, flowing 4,160 miles (6,693 km) northward from east central Africa into the Mediterranean Sea	The great, desolate expanse across north Africa that separates the Mediterranean Arab countries from the rest of Africa	The resource that is in short supply but is greatly needed for development of agricultural lands	The waterway at the western end of the Mediterranean Sea that separates Morocco from Spain by only 9 miles (14 km)
10	A fertile spot in a desert fed by an underground spring	The long body of water between Saudi Arabia and northeast Africa	The climate type of much of this region, where long, very hot summers force people to live along seacoasts and in river and mountain valleys	The area's most important resource, with the fields surrounding the Persian Gulf being the most productive in the world	The dam built across the Nile River as part of a huge irrigation project, which has greatly increased Egyptian agricultural production
15	A series of roughly parallel ranges located on the Mediterranean coasts of Morocco, Algeria, and Tunisia	An arm of the Arabian Sea, lying between Iran and the Arabian Peninsula	The term for the climate of the Atlas coast and parts of Turkey, Syria, Israel, and Lebanon that permits the raising of many fruits and vegetables	An expensive fuel generally found near oil in most of these countries, especially in Algeria and Iran	The salt lake on the border of Israel and Jordan whose surface lies 1,299 feet (396 m) below sea level

20

- The small peninsula between the Mediterranean and Red seas, belonging to Egypt, but part of the Asian continent
- The two rivers that rise in Turkey and flow through Iraq into the Persian Gulf, forming a wide plain
- The flooding river whose narrow banks of rich silt deposits support most of Egypt's agriculture
- The two essential minerals, needed for large-scale industrial production, that are lacking in this region
- A short, irregular river in the desert whose bed fills with a sudden rain but soon dries up again

25

- A very high range of snow-capped mountains that covers three fourths of Afghanistan
- The channel formed by the meeting of the Tigris and Euphrates rivers that flows into the Persian Gulf
- The name for the rich soil region that curves through the Iraqi plains, where several ancient civilizations developed
- The bark of trees grown in Morocco and Algeria that is valuable for its buoyancy, resistance to liquids, and soundproofing
- A poor Islamic nation in northwest Africa that is being devastated by the hot winds of the Sahara Desert

BONUS

- The northernmost country of this region—a mountain-rimmed plateau bordered on the north by the Black Sea
- The waterway that separates Turkey in Europe from Turkey in Asia and connects the Sea of Marmara with the Aegean Sea
- The desert covering a large portion of Saudi Arabia, which is the largest uninterrupted sand desert in the world
- An animal valuable as a commercial cleaning tool that is fished for in Mediterranean waters by Egypt, Tunisia, and Turkey
- The narrow strait controlled by the Turkish government at the outlet of the Black Sea at Istanbul

NOTES

Southwest Asia and North Africa

9

Topic: Occupations

	Agriculture	Mining	Manufacturing	Other Income	New Ways
5	A cash crop, grown mostly on plantations, of which Egypt grows some of the world's finest	A major producer of oil, and the country that exports more oil than any other country in the world	The industry common throughout the region, especially in Egypt, where it employs 40 percent of the nation's industrial workers	The animal raised throughout the region for meat and wool	One of the most industrialized countries in the Middle East, advanced by European ideas and technology
10	The fruits of the palm tree grown for food and export on many oases and in Iraq	The resource in such demand today that countries producing it have much influence on regional and world affairs	The business of producing articles at home, such as silverware, pottery, leather, and rugs	Nomadic tribes who wander in and near the deserts and mountains, herding sheep, goats, camels, and cattle	The small nation on the eastern shore of the Mediterranean Sea that is the most developed country in the Middle East
15	The grain that is the most important crop of the Middle East and a staple food of the people	A mineral used in fertilizers that is mined in large quantities in northwest Africa and in southern Israel	An important building material used to make concrete, and produced in Morocco and in countries in the eastern part of this region	A source of income in many places here, based on mild climate, historic ruins, interesting culture, and religious significance	The collective community of many Israeli farmers in which they share all property and combine their labor as equal partners

20

The main type of fruit grown in Israel and other Mediterranean lands, both as a food and as an export

A source of chromium, mined in Turkey, that is used to give steel hardness, shine, and rust resistance

A chemical salt taken from the Dead Sea that is used chiefly to make fertilizer

Valuable gems that are formed inside the shell of a kind of oyster fished for in the Persian Gulf

The food crop of the Nile Delta and other well-watered valleys, which has limits on cultivation because it uses too much water

25

The tree, valuable for its fruit and oil, that has been cultivated in the Mediterranean areas since ancient times

Mineral reserves found in the Atlas Mountains of Algeria that may help in the area's industrial development

Beautiful, expensive, hand-woven products for which Iran has long been world famous

The expensive delicacy made from the salted eggs of sturgeon caught in the Persian Gulf

The region in southern Israel that is now productive because an irrigation system brings water to it from Lake Tiberius (Sea of Galilee)

BONUS

The grain, hardier than wheat, that is used to feed livestock and to make malt beverages

The beautiful azure-blue mineral, used in ornaments, that is mined in Afghanistan

Foreign-owned plants that turn out finished products from imported parts, such as automobiles and motors

Revenue received from foreign oil companies that is used for modern development

A method of land reclamation used in Israel in which open hilly land is converted into forest by planting trees

NOTES

Southwest Asia and North Africa

9

Topic: People

PEOPLES	TRANSPORTATION AND TRADE	CITIES	PROBLEMS	POTPOURRI
The term for the common ancestry, language, religion, and culture that unites much of this region	An animal, sometimes called the ship of the desert, that carries goods and people over the shifting sands of the hot desert	The educational, cultural, and industrial center of Turkey, named Constantinople in ancient times	The country that both the Arabs and Israelis have claimed as their historic homeland, causing a long and bitter conflict	The main occupation of the people before oil was discovered in this region, and still the occupation of the majority
The small nation established by the United Nations in 1948 as a homeland for Jews from all over the world	The engineered waterway that separates Africa from Asia and connects the Mediterranean and Red seas	The capital of Iran, which has been modernized since 1925 with new streets, industries, and Persian-style government buildings	The process used in farming to make arid land productive	A phosphate-rich territory, now occupied by Morocco, that may choose to become independent
The followers of Islam, a major religion that has flourished in many diverse climatic, cultural, and ethnic regions of the world	The organization that has complete control of both the production and pricing of oil exports	The capital of Egypt and the largest city in Africa, influenced by ancient and modern cultures of the East and West	The strategically located country, landlocked and mountainous, that has been devastated by many years of strife and civil war	The capital of Iraq, an ancient city and famous market for Middle East wares, and now a modern center of government and industry

5

10

15

20

The city holy to Jews, Christians, and Muslims, which continues to be a source of conflict between Arabs and Israelis

The economical way of transporting oil from the oilfields to seaports for refining or shipping

Israel's largest metropolitan area, and an industrial and commercial center located on the Mediterranean Sea

A Mediterranean island country, whose Turkish-speaking Muslims have declared independence for their northern territory, severing ties with the Greek majority

The small desert country on the northwest coast of the Persian Gulf whose rich economy is based almost entirely on its petroleum products

25

The most holy city of the Muslims, whose chief industry is the care of pilgrims whose religion urges a visit to it at least once in a lifetime

The most famous transportation route through the mountains between Afghanistan and Pakistan, used by truck, donkey, and camel caravans

The capital of Saudi Arabia and the largest city on the Arabian Peninsula, built on the site of a oasis

The substance found in most Iranian lakes that is accumulating in the soil as a result of irrigation

The federation of seven Arab states, each ruled by a sheikh or emir

BONUS

Semi-nomadic tribes, consisting of non-Arab Muslims, who have fought for years to establish an autonomous territory in northern Iraq

An association of independent Arabic-speaking countries involved in political, economic, cultural, and social programs to benefit member states

Lebanon's capital city on the Mediterranean Sea, which is struggling to rebuild as a banking and commercial center after years of warfare and poverty

A situation prevalent throughout the region—except in Israel, Jordan, and Lebanon—that must be remedied before much advancement can occur

Crowded open-air trading and shopping centers in the old sections of southwest Asian cities

NOTES

UNIT 10

Sub-Saharan Africa

10 Sub-Saharan Africa

Topic: Physical Features

LAND	WATER	CLIMATE	NATURAL RESOURCES	POTPOURRI
5 A series of trenchlike depressions, created by a huge land shift, that stretch from Mozambique to the Red Sea	The river that crosses the equator twice as it makes a wide curve in its course to the Atlantic Ocean	The type of vegetation near the equator that results from hot, humid weather year-round	The condition of rain forest soil, which cannot support crops	The vast desert that separates the peoples of Africa and is an obstacle to interchange with Europe and north Africa
10 The landform that occupies most of the central part of Africa	The river that drains most of west Africa before it empties into the Gulf of Guinea through a huge delta	The area, north and south of the rain forest, that is covered with tall, coarse grass and occasional trees	Physical features abundant in Africa that are an excellent potential source of hydroelectric power	The spectacular waterfalls on the Zambezi River between Zambia and Zimbabwe
15 A large island republic off the southeast coast of Africa, extremely rich in natural wonders but poor in economic development	An inlet of the Atlantic Ocean in west Africa, south of the great bulge of the continent	A dry wasteland in Botswana and Namibia, parts of which can be used for grazing	The rich nation where long-lasting oil reserves and vast amounts of natural gas were discovered in the 1950s	Africa's highest mountain, which is covered with ice and snow year-round, although it lies near the equator

Storms that occur almost daily in the rain forest

A hot, dry, poor country whose economy is based on its well-located port on the Gulf of Aden

The animals that once were common across Africa but were killed by the thousands for their ivory tusks

The ore used to make aluminum foil, airplanes, and utensils, which is found in Guinea and Sierra Leone

A tough ore, found in Zambia and the Democratic Republic of the Congo, that resists high temperatures and is used for jet engines and pigments

An unusual African tree with an extremely wide trunk, whose leaves and bark are used for medicine, paper, cloth, and rope, and whose fruit is used for food

The grassland region, similar to the American prairie, that is situated on a high plateau in southeast Africa

The type of climate in the small area around Cape Town, which has hot, dry summers and mild, rainy winters, suitable for fruit raising

Desertification is causing this arid region of short grasses on the Sahara's southern border to rapidly shrink

Africa's largest lake and the source of the Nile River, lying between the east and west branches of the Great Rift Valley

A very long and deep freshwater lake that lies in the Great Rift Valley

The large strait between the island of Madagascar and the southeast African mainland

The country in east Africa that is mostly covered by high, rugged mountains

Very rugged mountains between the veld and coastal plain in South Africa, which form a shield against rain-bearing winds

The important river basin that lies in a lowland area near the equator

20

25

B O N U S

N O T E S

Sub-Saharan Africa

Topic: Occupations

10

	AGRICULTURE	FORESTRY	MINING	INDUSTRIES AND PRODUCTS	POTPOURRI
5	A versatile cash crop raised as a protein food and for its oil	The main crop of west Africa, where much of the world's supply is produced	A precious metal, one third of the world's supply of which is found in deep mines near Johannesburg in South Africa	The animal raised in large numbers in South Africa for its various export products	The type of farming done by the majority of Africans, who raise only enough food for their families
10	Herders who wander from place to place in dry savanna lands in search of water and food for their sheep, goats, and cattle	The tropical tree whose nutlike fruit yields an oil used for food and commercial products	A leading export of Zambia and the Democratic Republic of the Congo that is used to make pipes and electrical wire	The product derived from the palm and the peanut that is used to make cosmetics and machine lubricants	The dam built on the Zambezi River between Zambia and Zimbabwe, which supplies electric power for both nations
15	The leading crop of Sudan, whose production and irrigation is controlled by the government	Once set up by Europeans as a plantation crop, now grown mostly on small farms in the rain forest for export	The use made of the many diamonds found in South Africa near Pretoria and Kimberley	The product made from hides that is sold for export or used in the manufacture of shoes	Giant dam on the White Nile River near Khartoum, which was built for irrigation

An extension of land near Africa's southern tip whose sandy beaches make it a favorite vacation spot

Big-game hunting expeditions into Africa's grasslands

The disease-carrying insect that must be controlled before cattle can be raised throughout the savanna

The food product raised for export on tropical plantations in Mozambique, Mauritius, and Swaziland

An element needed for nuclear energy that is found and processed in Niger and South Africa

A crop grown on east African plantations and farms, whose leaf fibers are made into twine and rope

A prospering country that is the world's largest supplier of gem-quality diamonds

The nation that has one of the world's richest deposits of iron ore, on which it depends for half its income

A leading mineral of South Africa and Gabon that is used to make steel and chemicals

A hard, heavy, durable wood that is black and takes a high polish

A hard, reddish, tropical wood valuable for furniture making

A common savanna tree that produces gum arabic, a substance used in making glue, candy, and drugs

A main crop of Côte d'Ivoire (the Ivory Coast), Uganda, and Ethiopia that is used to make a beverage and medicines, and for export

An important grain in most Africans' diet that grows well where moisture is abundant

A staple grain, used for human and animal consumption, that is raised in hot, dry areas where no other grains can thrive

20

25

B O N U S

N O T E S

10 Sub-Saharan Africa

Topic: People

	PEOPLES	TRANSPORTATION AND TRADE	CITIES	PROBLEMS	SOLUTIONS
5	The predominant culture and race of sub-Saharan Africa	The obstructions that prevent Africa's rivers from being navigable	One of South Africa's capitals and its largest city, whose excellent harbor has made it an important shipping and trading center	The most developed and powerful of all sub-Saharan African countries, which has recently made great gains in civil rights	The goal all African nations have achieved since 1960, either by fighting or by peaceful agreement
10	West Africa's wealthiest and most populous nation	The chief means of transportation by which goods are brought from the interior to the coast	Kenya's capital and one of the largest and fastest-growing cities in Africa, this city is an economic, cultural, and tourist center	Central Africa's largest nation in both population and land area—a nation which has great economic potential but remains impoverished due to government corruption	A new source of income in Kenya and Tanzania where travelers experience natural wonders while wildlife and wildlands remain unaltered
15	The term used for ancient societies of black people with special territories, customs, and traditions	The river that, with its tributary the Ubangi River, serves as a major transportation route into central Africa	A city in South Africa in the heart of the world's richest gold field and near coal and iron deposits	The source of dispute among many new nations that arose from the way Europeans split up tribal lands	The country in east Africa, once British, whose tremendous progress through native and European cooperation is now threatened by government corruption

The resource that is especially protected in national parks and preserves

The southwest African country on the Atlantic coast that is largely desert but rich in mineral deposits

The association of independent African nations formed to promote unity and cooperation and to remove all forms of colonialism

The need of the newly independent countries in order to solve their social, economic, and political problems

The situation in many African countries that has caused much bloodshed, poverty, famine, disease, and economic collapse

The huge obstacle to economic recovery that exists in most of these countries

The capital, largest city, and chief marketplace of Ethiopia, as well as the headquarters of the Organization of African Unity (OAU)

The westernmost city in Africa, whose location on Cap Vert (Cape Verde) Peninsula makes it an important transportation and commercial center

The former capital of Tanzania and a major port in southeast Africa, connected with Zambia by railroad

The situation of both Zambia and Zimbabwe that is a disadvantage to them in exporting their farm and mineral products

The capital of the Republic of the Congo, and the eastern terminus of the Congo-Ocean Railroad that connects the Congo River with the port of Pointe-Noire on the Atlantic Ocean

The capital of Sudan, located at the junction of the Blue Nile and the White Nile, and a trading and communication center

The two languages mostly used in the schools, unifying people who also speak an immense variety of tribal languages

The small republic in west Africa that was established as a homeland for liberated slaves from the United States

The name given to the largest group of black Africans and their language

20

25

B O N U S

N O T E S

UNIT 11

Australia and Oceania

Australia and Oceania
Topic: Physical Features

11

	LAND	WATER	ISLANDS	CLIMATE	NATURAL RESOURCES
5	Mountain ranges in eastern Australia that are not high but rise steeply from the coastal plain	The river in southeastern Australia that, with its tributaries, irrigates a vast area	The country lying 1,200 miles (1,932 km) southeast of Australia that consists of two large islands and a number of very small islets	An arid region in the center of Australia to which these names refer: Great Sandy, Gibson, and Great Victoria	Wells that use natural water pressure and supply water to much of Australia's grazing land
10	The landform that occupies three fifths of Australia, broken only by a few hilly ranges	Australia's longest river, which rises in the eastern mountains and flows southwest into the Murray River	The Australian island state located south of the continent	A tropical storm in the Pacific that causes serious damage with its violent winds and torrential rains	Dried coconut meat, which yields an oil used for soap, candles, and margarine, and which is the chief export of most Pacific islands
15	The range that runs the whole length of New Zealand's South Island and contains snowfields and glaciers	That part of the South Pacific Ocean between southeast Australia and western New Zealand	The country that occupies the eastern half of the island of New Guinea	The climate type of northern Australia and most of the Pacific islands	The power produced from New Zealand's many swift-flowing rivers, supplying most of the country's needs

20	Australia's northernmost point, important for bauxite and as an Aborigine reserve	The wide, shallow bay that indents into the southern coast of Australia	Limestone from the skeletons of small sea animals that forms the upper portion of low islands	The climate type of New Zealand, where ocean breezes bring mild, moist weather year-round	The atomic energy source discovered in Australia's Northern Territory near Darwin and in other areas

25	The landform east of the mountains that has the best farming land and the largest cities in Australia	A large, shallow salt lake in south central Australia, or a vast expanse of dry salt, depending on the season	The kind of islands that are formed by built-up volcanoes and that are usually mountainous and fertile	The climate in southwest Australia and the area east of the Great Australian Bight	The unusual source of electricity on the volcanic plateau of New Zealand's North Island

B O N U S	The largest of Australia's underground basins, which produces water that is too salty for human use or irrigation, but is good for watering livestock	Water within an atoll	A ring-shaped coral reef, or low island, that surrounds a lagoon	Winds that bring abundant rainfall to Tasmania, New Zealand's west side, and the southernmost part of Australia	The term that applies to New Zealand's flightless birds and fuzzy green fruit

N O T E S

Australia and Oceania

Topic: Occupations

11

AGRICULTURE	FORESTRY AND FISHING	MINING	INDUSTRIES AND PRODUCTS	OTHER INCOME
5 Australia's chief crop, grown for food, export, and fodder, often raised on the same farms as sheep	The plantation tree raised for food, copra, and oil—the most important products of the Pacific islands	An important mineral, both black and brown, that is found in eastern Australia and used in the country's manufacturing industries	The leading product of Australia and New Zealand, with which they supply one third of the world's market	The main occupation of most of the Pacific islands
10 The animals raised for food on huge stations, or ranches, in Australia's outback and eastern region	The largest coral reef in the world, along Australia's northeast coast, which is a threat to ships but a favorite spot for deep-sea divers	Two of Australia's leading metals that are often found together—one used for pipes, the other for galvanizing, and both for paint and batteries	The two meats obtained from the dual-purpose sheep raised in the humid parts of Australia and New Zealand	The source of cash income promoted by some islands but considered by others as a threat to nature and native customs
15 The hardy breed of sheep raised in Australia and New Zealand that produces a high-quality wool	An occupation of people living in eastern coastal regions that supplies food mostly for home consumption	The ore of which Australia is the world's leading producer, with reserves lying in the southwest and in the Cape York Peninsula	Foods produced in highly mechanized plants as major export products as well as for a large home market	The name for money that is brought in from another country, through exports, foreign tourists, or wages from citizens working abroad

20

Large animals raised in Australia and New Zealand, where abundant rains make good pasturelands

The country whose wood products come mainly from its large commercial forests of radiata pine

An important mineral found near Whyalla on Australia's Spencer Gulf and also in the northwestern part of the continent

Industries using iron ore and coal at Whyalla, Brisbane, and Newcastle

The establishments built by Western nations on the strategically located Pacific islands, which still provide a cash income for the islanders

25

Australia's most important fruit crop grown in the temperate regions

A gum tree that produces hard timber and oils for medicine and industrial uses

The metal that brought Europeans to Australia in the 1800s, now mined at Kalgoorlie and in Papua New Guinea

A tropical plantation crop grown on the islands, in New Zealand, and on the humid eastern coast of Australia, where it is refined

A famous mine in Queensland, Australia, that is among the world's largest for mining lead, silver, zinc, and copper

BONUS

A fruit grown in the irrigated Murray River Basin and Tasmania, used to make wine

A very versatile tree of the tropical islands, providing food, building materials, and other essential items

The island that is a leader in the mining and smelting of nickel and that has deposits of chromite as well

Manufacturing plants, mostly foreign-owned, that use steel to produce Australia's major means of travel

The nation in which foreign interests have developed one of the world's largest copper mines

NOTES

Australia and Oceania

Topic: People

11

	PEOPLES	TRANSPORTATION AND TRADE	CITIES	PROBLEMS AND SOLUTIONS	POTPOURRI
5	The official language of Australia and New Zealand, and the one most widely used in the Pacific islands	The means of transport that enables Australia and New Zealand to export many meat and dairy products	Australia's oldest and largest city, which lies on the southeast coast on one of the finest harbors in the world	The goal that some Pacific islands have already achieved (since 1960) and that most of the other islands are working toward	The month that has the first day of summer in this region
10	The earliest Australians, who are fighting to receive recognition and reparations from the government for centuries of mistreatment	An excellent transportation system that makes most of Australia's interior areas accessible	A large city and port in southern Australia, whose industries include shipbuilding, oil refining, and manufacturing consumer goods	The area from which most Australian immigrants come, supplementing the workforce	The 180° meridian that determines the instant one day ends and the other begins and that separates Polynesia from the West
15	The widely scattered group of islands in the southeast Pacific Ocean which includes Samoa, Tuvalu, Tahiti, and the Cook Islands	The country that imports a large part of Australia's minerals as raw materials for industries	A modern city in southwest Australia, blessed with an ideal climate and linked to eastern cities by railroads	The island group that has remained the most isolated because of its thick jungles, tall mountains, and tropical diseases	The parallel of 23°30' south latitude that passes through central Australia and south of most of the Pacific islands

The group of more than 600 islands northeast of New Guinea, which include Palau, Yap, Kiribati, and the Marshall Islands	Once Australia's and New Zealand's principal export market, this country now does much business with the European Union	An important port and market on Australia's east coast, whose chief manufactured products include ships and processed food
The island group in the west Pacific Ocean which includes Papua New Guinea, Fiji, Vanuatu, New Caledonia, and the Solomon Islands	The country called the Crossroads of the South Pacific because it lies on major shipping routes and has excellent harbors	New Zealand's largest city and manufacturing center, located on an excellent harbor on North Island
The collective name used for most of the islands in the Pacific Ocean, whose subdivisions are grouped according to the physical and cultural traits of the inhabitants	The popular mode of transport that changes in each state but now connects Sydney and Perth	The small, carefully planned city in the Australian capital territory that serves as Australia's center of government

The independent nation that prohibits nuclear-armed and nuclear-powered vessels from using its port facilities	An independent nation, composed of 607 islands, formerly known as the Caroline Islands	The small animal pest that eats much of the grazing grass and must be controlled by the government

A dry lake in the interior that usually contains little or no water	The term for the vast remote arid region of central and western Australia	A native marsupial of the dry grasslands that can be a nuisance to stock raisers and motorists

20

25

B O N U S

N O T E S

UNIT 12

Polar Regions

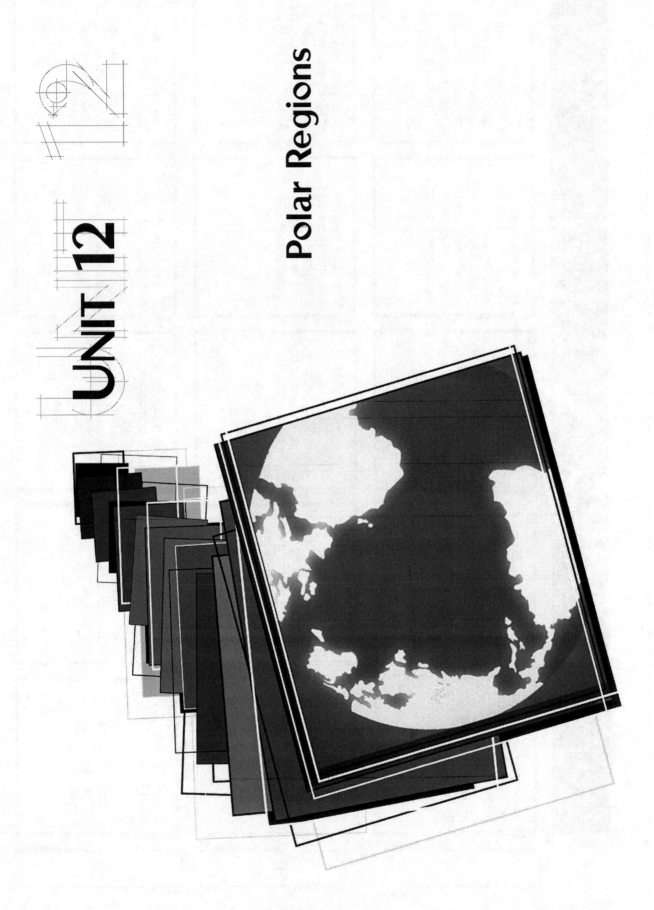

12 Polar Regions
Topic: Natural Features

	LAND AND WATER	CLIMATE AND NATURAL RESOURCES	OCCUPATIONS	TRAVEL	POTPOURRI
5	The narrow waterway between the United States and Russia that connects the Arctic Ocean with the Pacific	Plains north of the tree line underlaid with permafrost, where mosses, lichens, and grass grow in the short summer	The chief occupation of the Arctic, due to the excellent continental shelf areas off Greenland, Iceland, Norway, and Russia	Huge, floating chunks of ice that have broken from glaciers or ice shelves and are a major threat to polar ocean travel	A temporary Inuit (Eskimo) hunting camp—a dome-shaped hut of hard snow for winter, a tent of skins for summer
10	Floating masses of piled-up ice that cover most of the Arctic Ocean year-round	The phenomenon that results when the earth's axis tilts so that one pole receives constant sun for six months while the other remains in darkness	The animal, valuable for meat and oil, that is caught in Antarctic waters, according to strict international regulations	The chief means of hauling game and equipment over ice and snow in the Arctic region	Stocky marine birds of black-and-white plumage that live in gigantic colonies on Antarctic coastal ice
15	A long, narrow, mountainous peninsula in Antarctica, reaching within 600 miles (965 km) of South America	The line of latitude that determines the Arctic region in the north and the Antarctic region in the south	The purpose for which numerous countries have established settlements in Antarctica, demonstrating international cooperation	The means of transport that delivers or drops supplies where ships cannot go	The name for the hardy Arctic people, once nomadic hunters, many of whom now live in fixed settlements

The mammal of cold coastal regions, valuable for its fur, hide, and oil, and hunted only under international restrictions

A powerful ship equipped to clear a channel through ice fields for supply and cargo ships

The compromise signed in 1991 that bans mineral and oil exploration on Antarctica for a half-century and that specifies the region to be used for peaceful purposes only

A water passage through the Arctic Ocean north of Europe and Asia that, after being cleared by icebreakers, is used by many vessels each summer

Shrimplike creatures that form the staple diet of whales and are now being developed for human consumption

The method used in the Arctic to catch animals for food and valuable furs

Isolated Arctic facilities where scientists study and predict storms and conditions of areas farther south

A water route from the Bering Strait across Canada that is only used by a few ships carrying supplies to weather and radar stations

The factor that, in combination with extremely dry, cold air, makes Antarctica the harshest, most desolate place on the earth

Small plants and animals that live in polar waters, providing food for fish and larger animals

A source of energy for the United States, discovered on Alaska's North Slope in the late 1960s

The sea forming a deep indentation into Antarctica, which is partially filled with an ice shelf of the same name

Mountains that cross the entire continent of Antarctica, dividing it into West Antarctica and the larger East Antarctica

The mineral mined in Alaska, Canada and its northern islands, and Russia, and found in good supply in Antarctica

The ocean current that flows from the Arctic Ocean, bringing ice to the east coast of Greenland, and that becomes the Labrador Current

20

25

B O N U S

N O T E S

UNIT 13

World Perspectives

13

World Perspectives
Topic: Natural Features

	LATITUDES	CLIMATE	DIRECTIONS	LAND	WATER
5	The latitude at which daylight and nighttime are equal, with the sun rising and setting at the same time every day of the year	The two continents on which there are no tropical rain forests	The continent on which lie the countries east of the Suez Canal	The only country that is also a continent, the world's largest island, and the smallest and flattest continent	The one continuous body of salt water that covers three fourths of the earth and plays a vital role in the climate of the earth
10	The only Asian country through which the equator passes	The continent that receives the least amount of precipitation	The direction in which you would travel from Miami, Florida, U.S.A., to Quito, Ecuador	The country that serves as a land link between North and South America, as well as a sea link between the Atlantic and Pacific oceans	The waterway that made it possible to sail from Europe to Asia without rounding the southern tip of Africa
15	The continent, other than Antarctica, that extends farthest toward the South Pole	The directional name given to a wind blowing from eastern Canada to New York and the New England states	The island group that lies east of Vietnam	The largest island in Europe	The world's largest lake, salty because it is completely landlocked

The line of latitude that passes through the Sahara Desert

The warm, surface ocean current that carries warm, moist air from the Caribbean to the British Isles and Scandinavia, causing mild winters

The direction in which you would travel from Japan to North Korea

The country that is 2,650 miles (4,260 km) long (about as far as New York to San Francisco) and only 250 miles (400 km) wide at its greatest width

The bay bordered by India, Bangladesh, and Myanmar

20

A shimmering, colorful light display seen in the night sky around the geomagnetic poles, named borealis in the north, australis in the south

A high-speed wind that moves eastward at high altitudes and that occurs where large temperature differences exist in the atmosphere

The direction in which the North Sea lies from Great Britain

The third country, along with Sweden and Russia, that is adjacent to Finland's borders

The body of water into which the Yukon River empties

25

The twice-yearly occurrence when the sun appears directly overhead at the equator, resulting in nearly equal periods of daylight and darkness worldwide

The equivalent temperature for how cold the air feels when the wind blows at a certain speed in a certain air temperature

The direction you would face if you were standing at the North Pole (90° north latitude)

The massive underwater mountain system, 12,000 miles (19,308 km) long, that runs down the middle of the Atlantic Ocean

The deepest and oldest lake on the earth, which also holds the most water

B O N U S

N O T E S

13 World Perspectives
Topic: People

	POPULATION CHANGES	ENVIRONMENT	PROBLEMS	SOLUTIONS	POTPOURRI
5	The study of population and related subjects—vital statistics, growth, density, and distribution	The resource whose destruction is causing changes in weather patterns, soil erosion, water pollution, and the loss of native species and indigenous culture and knowledge	The human disaster that may result from the land's failure to produce adequate food, from economic instability, or from political strife	The process of change measured by a country's mechanized industry, technology, systems of transportation and communication, education, and life expectancy	The term that refers to people, plants, and animals that occur naturally in a certain area
10	The movement of people from one place to another	Precipitation mixed with harmful gases from exhaust and factories, which is ruining forests, lakes, wildlife, and even architecture	The necessary ingredient for growth that many developing countries cannot attain due to tribal and ethnic loyalties	The extension of economic activity sought by many developing countries that have been overly dependent on a single cash crop or commodity	Fuels composed of hydrocarbons, formed from the remains of plants and animals that lived millions of years ago—coal, oil, and natural gas
15	The collective term for the industries employing workers in these fields—finance, medicine, education, trade, transportation, communication, utilities, and government	The layer of gas in the stratosphere that absorbs most of the sun's harmful ultraviolet radiation, and that is threatened by certain manufactured chemicals	The encroachment of sand onto adjoining grasslands because of overgrazing, overuse of water, or unwise use of land	The program by which cities try to bring back families and businesses in order to overcome poverty, crime, and unemployment	The belt around the Pacific Ocean that consists of hundreds of active volcanoes

The only city, in the same country, that straddles the European and Asian continents

Shock waves through solid rock generated by earthquakes or underground explosions

The total market value of all the goods and services produced by a country in a year

The ability to read and write, and a measure of a country's level of development

The removal of salts and other chemicals from seawater—a costly way of providing necessary water, especially in Africa

The method by which fields are top-tilled to protect topsoil from water and wind erosion, saving water, soil, and money

The global food source that is becoming depleted, resulting in unemployment, food shortages, and ecological imbalance, which is extremely difficult to reverse

The rapid clearing of trees for firewood, farming, and grazing that can harm the environment by affecting rainfall, temperature, soil conditions, and plant and animal life

The illegal capture or killing of wild animals, making conservation difficult

The increase of carbon dioxide in the atmosphere from the burning of fossil fuels and the destruction of forests, resulting in the steady, gradual rise in the temperature of the earth

Development that uses natural resources without destroying the basis of their productivity, allowing them to regenerate

The industry that invites tourists to experience natural wonders and generate local revenues while preserving what they have come to see

People who flee from danger, poverty, famine, or persecution, usually to places of security and economic opportunity

A region made up of several large cities and their surrounding areas in close enough proximity to be considered a single urban complex

The term once used for underdeveloped countries or those needing major economic and social improvements

20

25

B O N U S

N O T E S

Answer Key

1 Geographical Terms and Understandings
Topic: Maps

	Basic Terms	Latitude	Longitude	Kinds of Maps	Map Reading
5	What is a hemisphere?	What is the equator?	What is an axis?	What is a globe?	What is a legend?
10	What is longitude?	What is the Tropic of Cancer?	What is the prime meridian?	What is a physical map?	What is a symbol?
15	What is latitude?	What is the Tropic of Capricorn?	What is the international date line?	What is a political map?	What is scale?
20	What is a meridian?	What is the Arctic Circle?	What are time zones?	What is a topographic map?	What is a grid?
25	What is a parallel?	What is the Antarctic Circle?	What is west?	What is a relief map?	What are contour lines?
BONUS	What is cartography?	What is a pole?	What is daylight saving time?	What is the Mercator projection?	What are isotherms?

1 Geographical Terms and Understandings
Topic: Land

	Basic Terms	Landforms	Elevations	Change Processes	Potpourri
5	What is a continent?	What is an island?	What is a mountain?	What is an earthquake?	What is the continental shelf?
10	What is sea level?	What is a peninsula?	What is a hill?	What is a volcano?	What is the continental divide?
15	What is altitude?	What is a cape?	What is a valley?	What is a fault?	What is the fall line?
20	What is a coast?	What is an isthmus?	What is a plateau?	What is a drought?	What is a basin?
25	What is a subcontinent?	What is an archipelago?	What is a plain?	What is reclaiming?	What is a range?
BONUS	What is landlocked?	What is an atoll?	What is a canyon?	What is coral?	What is a reef?

1 Geographical Terms and Understandings
Topic: Water

	Basic Terms	Water Formations	Rivers	Water on Land	Potpourri
5	What is an ocean?	What is a lake?	What is a source?	What is a glacier?	What is a canal?
10	What is a river?	What is a gulf?	What is a mouth?	What is a wetland?	What is a reservoir?
15	What is the water cycle?	What is a bay?	What is a tributary?	What is groundwater?	What are rapids?
20	What is a current?	What is a strait?	What is a delta?	What is a wadi?	What is a lagoon?
25	What are tides?	What is a fjord?	What is an estuary?	What is a playa?	What is a tsunami?
B O N U S	What is oceanography?	What is an inlet?	What is a levee?	What is the water table?	What is an iceberg?

1 Geographical Terms and Understandings
Topic: Climate and Weather

	Basic Terms	Climate Types	Winds	Ocean Currents	Storms
5	What is weather?	What is desert?	What is windward?	What is the Gulf Stream?	What is a thunderstorm?
10	What is climate?	What is tropical?	What is leeward?	What is the Labrador Current?	What is a tornado?
15	What is atmosphere?	What is continental?	What are prevailing winds?	What is the North Atlantic Drift (or Current)?	What is a hurricane?
20	What is precipitation?	What is Mediterranean?	What are trade winds?	What is the Peru (or Humboldt) Current?	What is a monsoon?
25	What is humidity?	What is marine?	What are doldrums?	What is the Japan Current?	What is a typhoon?
B O N U S	What is meteorology?	What is tundra?	What is air pressure?	What is the Brazil Current?	What is a front?

1 Geographical Terms and Understandings
Topic: Soil and Vegetation

	Basic Terms	Soil	Vegetation Regions	Change Processes	Plant Potpourri
5	What is environment?	What is humus?	What is a rain forest?	What is erosion?	What is a growing season?
10	What is conservation?	What is permafrost?	What is a desert?	What is leaching?	What is a tree (or timber) line?
15	What is an oasis?	What is arable?	What is a tundra?	What is fertilizing?	What is evergreen?
20	What is rain shadow?	What is alluvial?	What is a savanna?	What is irrigating?	What is deciduous?
25	What is ecology?	What is silt?	What is a steppe?	What is a floodplain?	What is dormant?
BONUS	What is geology?	What is loam?	What is a taiga?	What is weathering?	What is scrub?

2 The United States
Topic: Northeastern States

	Land and Water	Climate and Resources	Land Use	Industries	People and Places
5	What is the Chesapeake Bay?	What is the Atlantic Coastal Plain?	What is Pittsburgh (Pennsylvania)?	What is Washington, DC?	What is New York City (New York)?
10	What is the Ohio River?	What is hydroelectric power?	What is dairy farming?	What is Wall Street?	What is Philadelphia (Pennsylvania)?
15	What is the fall line?	What is anthracite?	What is Chesapeake Bay?	What is insurance?	What is Boston (Massachusetts)?
20	What is the Piedmont?	What is Niagara Falls?	What is lumber?	What is New England?	What is a megalopolis?
25	What is the Connecticut River?	What is the (sugar) maple?	What is tourism?	What is electronic equipment?	What are tax revenues?
BONUS	What is New England?	What is marble?	What is the potato?	What is Wilmington (Delaware)?	What is the United Nations?

2 The United States
Topic: North Central States

	Land and Water	Climate and Resources	Land Use	Industries	People and Places
5	What are the Central Plains?	What is winter wheat?	What is the Corn Belt?	What is Minnesota?	What is Chicago (Illinois)?
10	What is the Mississippi River?	What is a dust storm?	What is wheat?	What is meatpacking?	What is Detroit (Michigan)?
15	What is Lake Michigan?	What is a tornado?	What is Wisconsin?	What is flour milling?	What is Kansas City?
20	What is the Missouri River?	What is coal?	What are grain elevators?	What is the Calumet Region?	What is St. Louis (Missouri)?
25	What is the St. Lawrence Seaway?	What are Illinois and Iowa?	What are soybeans?	What is Akron (Ohio)?	What is Cleveland (Ohio)?
BONUS	What are the Black Hills?	What is the ice sheet (or glacier)?	What are cooperatives?	What is Wichita (Kansas)?	What is Milwaukee (Wisconsin)?

2 The United States
Topic: Southern States

	Land and Water	Climate and Resources	Land Use	Industries	People and Places
5	What is the Mississippi River?	What is coal?	What is the Cotton Belt?	What is oil refining?	What is New Orleans (Louisiana)?
10	What is the Continental Shelf?	What is citrus?	What is tobacco?	What is poultry?	What are Dallas and Fort Worth (Texas)?
15	What is the Piedmont?	What is natural gas?	What is lumber?	What are petrochemicals?	What is Miami (Florida)?
20	What are the Ozark Mountains?	What are sulfur and salt?	What is livestock raising?	What is Cape Canaveral (Florida)?	What is Appalachia?
25	What is the Edwards Plateau?	What is bauxite?	What is the peanut?	What is Birmingham (Alabama)?	What is Atlanta (Georgia)?
BONUS	What are levees?	What is shrimp?	What is the Tennessee Valley Authority (TVA)?	What is coke (or coking coal)?	What is Houston (Texas)?

2 The United States
Topic: Western States

	Land and Water	Climate and Resources	Land Use	Industries	People and Places
5	What is the Continental Divide?	What is the Colorado River?	What is irrigation?	What is California?	What is Los Angeles (California)?
10	What is the Central Valley?	What is the Columbia River?	What is lumbering?	What is Hawaii?	What is San Francisco (California)?
15	What is Oahu?	What is oil shale?	What is salmon?	What is New Mexico?	What is Seattle (Washington)?
20	What is the Grand Canyon?	What is Mediterranean?	What is Wyoming?	What is Nevada?	What is Alaska?
25	What are the Aleutian Islands?	What is smog?	What is copper?	What is the federal (or U.S.) government?	What is Denver (Colorado)?
B O N U S	What is the Great Salt Lake?	What is Yellowstone?	What is pineapple?	What is Valdez (Alaska)?	What is skiing (or winter sports)?

3 Canada
Topic: Physical Features

	Land	Water	Climate and Soil	Natural Features	People and Places
5	What is the Canadian Shield (or Laurentian Plateau)?	What are the Great Lakes?	What is the tundra?	What are the Grand Banks?	What are Inuits (or Eskimos)?
10	What are the Canadian Rockies?	What is Hudson Bay?	What is the Canadian Shield?	What is hydroelectricity?	What is Québec?
15	What is the Northwest Territories?	What is Niagara Falls?	What are beef cattle?	What is (fresh) water?	What is acid rain?
20	What is Newfoundland?	What is the Bay of Fundy?	What is marine?	What is uranium?	What is Nunavut?
25	What is Maritime?	What is Mackenzie?	What is the Lakes Peninsula (and St. Lawrence Valley)?	What is salmon?	What is mining?
B O N U S	What is Ellesmere Island?	What is the Fraser River?	What is permafrost?	What is Churchill Falls?	What is the North American Free Trade Agreement (NAFTA)?

3 Canada
Topic: Occupations

	Farming and Forestry	Mining	Industries and Products	Transportation	Cities
5	What is wheat?	What is nickel?	What is tourism?	What is the St. Lawrence Seaway?	What is Montreal?
10	What is timber?	What is silver?	What are motor vehicles and parts?	What are the Sault Sainte Marie (or Soo) Canals?	What is Toronto?
15	What is Prairie?	What is iron ore?	What is steel?	What is the Welland (Ship) Canal?	What is Vancouver?
20	What are pulp and paper (or newsprint)?	What is Cape Breton Island?	What is potash?	What is Winnipeg?	What is Ottawa?
25	What is dairy?	What is zinc?	What are service industries?	What is Halifax?	What is Québec?
BONUS	What is Prince Edward Island?	What are bituminous (or oil, or tar) sands?	What is Kitimat?	What is the Trans-Canada Highway?	What is Thunder Bay?

4 Middle America and the Caribbean
Topic: Physical Features

	Land	Water	Climate and Soil	Natural Resources	Potpourri
5	What is Cuba?	What is the Caribbean Sea?	What are the West Indies?	What is Mexico?	What is the Panama Canal?
10	What are Sierra Madre Oriental and Sierra Madre Occidental?	What is the Rio Grande?	What is (high) altitude?	What is hydroelectricity?	What is Hispaniola?
15	What is Baja California?	What is the Gulf of California?	What is rain forest?	What is natural gas?	What is Puerto Rico?
20	What is the Yucatán Peninsula?	What is the Gulf of Mexico?	What are hurricanes?	What is chicle?	What is the Central Highlands (or Plateau of Mexico, or Central Plateau)?
25	What are the Bahama Islands (or Bahamas)?	What is Gatun Lake?	What are earthquakes?	What is sisal?	What is the Netherlands?
BONUS	What is Bermuda?	What is Lake Nicaragua?	What is cattle grazing (or raising)?	What is ebony?	What is (a) Creole?

4 Middle America and the Caribbean
Topic: Occupations

	Agriculture	Mining	Manufacturing	Tourism	Government Influence
5	What is corn (or maize)?	What is Trinidad and Tobago?	What is Monterrey?	What is warm climate (or warm winters)?	What is Haiti?
10	What are bananas?	What is Mexico?	What is sugar?	What is ecotourism?	What is Trinidad and Tobago?
15	What is coffee?	What is Jamaica?	What are assembly plants?	What is Belize?	What is civil war?
20	What is sugarcane?	What is sulfur?	What is Puerto Rico?	What are handicrafts?	What is industry?
25	What is cotton?	What is fluorite (or fluorspar)?	What is mahogany?	What are ruins?	What is irrigation?
B O N U S	What is Grenada?	What is Trinidad and Tobago?	What is tobacco?	What are the United States Virgin Islands?	What is the United States?

4 Middle America and the Caribbean
Topic: People

	Peoples	Cities	Problems	Solutions	Changes
5	What is a mulatto?	What is Mexico City?	What is high population growth?	What is the Organization of American States (or OAS)?	What is Panama?
10	What is a mestizo (or Latino)?	What is Havana?	What are primitive (farming) methods?	What is emigration?	What is Cuba?
15	What is black?	What is San Juan?	What is Mexico City?	What is the Pan-American Highway?	What is Nicaragua?
20	What is Spanish?	What is Panama City?	What are schools?	What is Costa Rica?	What is El Salvador?
25	What is subsistence farming?	What is Costa Rica?	What are shantytowns (or slums)?	What is Ejido?	What is Honduras?
B O N U S	What is Guatemala?	What is Guatemala City?	What is unemployment?	What is the North American Free Trade Agreement (or NAFTA)?	What is Aruba?

5 South America
Topic: Physical Features

	Land	Water	Climate	Natural Resources	Potpourri
5	What are the Andes Mountains?	What is the Amazon River?	What is the Atacama Desert?	What is hydroelectricity?	What is June?
10	What are the Brazilian Highlands?	What is the Río de la Plata?	What is the Peru (or Humboldt) Current?	What is gold?	What are earthquakes?
15	What are the Guiana Highlands?	What is Lake Maracaibo?	What are pampas?	What is the rain forest?	What is east?
20	What is the altiplano?	What is Lake Titicaca?	What is Patagonia?	What are the Galápagos Islands?	What is oxygen?
25	What is Tierra del Fuego?	What is the Orinoco River?	What is savanna?	What is leached?	What is the gran chaco?
BONUS	What is Mount Aconcagua?	What is the Strait of Magellan?	What is altitude?	What is a llama (or alpaca)?	What is Angel Falls?

5 South America
Topic: Occupations

	Agriculture	Fishing and Forestry	Mining	Manufacturing	More Income
5	What is sheep raising?	What is Peru?	What is Venezuela?	What are textiles?	What is coca?
10	What is coffee?	What is cacao?	What is Chile?	What is wheat?	What is wool?
15	What are bananas?	What is mahogany?	What is Bolivia?	What is bauxite?	What are llanos?
20	What are cattle?	What is rubber?	What is iron?	What is (refined) sugar?	What is Uruguay?
25	What is cotton?	What is the Brazil nut?	What is Peru?	What is Chile?	What is the Organization of Petroleum Exporting Countries (or OPEC)?
BONUS	What is subsistence?	What is weight?	What are emeralds?	What is a labor force?	What is agriculture (farming)?

5 South America
Topic: People

	Peoples	Cities	Problems	Change	Closing the Gap
5	What is a mestizo (or Latino)?	What is São Paulo?	What is the rain forest?	What is French Guiana?	What is the Organization of American States (or O.A.S.)?
10	What is white?	What is Buenos Aires?	What is drug trafficking (or cocaine trade)?	What is Guyana?	What is the Pan-American Highway?
15	What is black?	What is Caracas?	What are slums (or shantytowns)?	What is Suriname?	What is isolation?
20	What is Native American (or indigenous peoples)?	What is Bogotá?	What is illiteracy?	What is money?	What is air transport?
25	What is Brazil?	What is Rio de Janeiro?	What is food production?	What is revolution?	What is Chile?
BONUS	What is Spanish?	What is Brasília?	What are Bolivia and Paraguay?	What is Paraguay?	What is Argentina?

6 Europe
Physical Features

	Land	Water	Climate and Soil	Natural Resources	Potpourri
5	What is the Alpine mountain system?	What is the Rhine River?	What is Mediterranean?	What is (Upper and Lower) Silesia?	What is Scandinavia?
10	What are the Pyrenees?	What is the Danube River?	What is the Gulf Stream (or North Atlantic Drift or Current)?	What is hydroelectricity?	What is a polder?
15	What is Norway?	What is the Baltic Sea?	What is the North (or Great) European Plain?	What is Reykjavík?	What is Venice?
20	What is Greece?	What is the Adriatic Sea?	What is the Meseta?	What is lignite?	What is the North Sea?
25	What are the Apennines?	What is the Po River?	What is Sicily?	What is nuclear?	What is Greenland (Kalaallit Nunaat)?
BONUS	What is Denmark?	What is Finland?	What is loess?	What is peat?	What is the reindeer?

6 Europe
Topic: Occupations

	Agriculture	Forestry and Fishing	Mining	Manufacturing	More Income
5	What is wheat?	What is lumber?	What is coal?	What are woolen goods?	What is Switzerland?
10	What is wine?	What is Norway?	What is the Ruhr (River Valley)?	What is luxury?	What is Denmark?
15	What are sugar beets?	What is winter?	What is iron ore?	What is Malta?	What is the Riviera (or Côte d'Azur)?
20	What is the potato?	What is Iceland?	What is the North Sea?	What is flax?	What is Ireland (Eire)?
25	What is rye?	What is Portugal?	What is Belgium?	What is Italy?	What is dairy farming?
B O N U S	What is the olive?	What are whales?	What is Sweden?	What is coal?	What are flower (or tulip) bulbs?

6 Europe
Topic: People

	Peoples	Cities	Transportation and Trade	Governments	Potpourri
5	What is Northern Ireland?	What is London?	What is the European Union (or EU)?	What is the United Kingdom?	What is England?
10	What is Vatican City?	What is Paris?	What is the Eurotunnel (or Chunnel)?	What is Germany?	What is pollution?
15	What is Monaco?	What is Rome?	What is Rotterdam?	What is Yugoslavia?	What are cooperatives?
20	What is Lapland?	What is Athens?	What are merchant fleets?	What is Berlin?	What is neutral?
25	What are migrant workers?	What is Copenhagen?	What are canals?	What are the Czech Republic and Slovakia?	What is Poland?
B O N U S	What is ethnic?	What is Vienna?	What is the Midland (Mitteland) Canal?	What is Gibraltar?	What is Albania?

7 Northern Eurasia
Topic: Physical Features

	Land	Water	Climate and Soil	Cities	Interesting Facts
5	What is Ural?	What is the Volga River?	What is the tundra?	What is Moscow?	What is the Commonwealth of Independent States (C.I.S.)?
10	What are the Caucasus Mountains?	What is the Caspian Sea?	What is the steppe?	What is Tashkent?	What is Russia?
15	What is Siberia?	What is the Black Sea?	What is desert?	What is Vladivostok?	What are Latvia, Estonia, and Lithuania?
20	What is a plain?	What is the Ob River?	What is the taiga?	What is Rostov (-on-Don)?	What is the Aral Sea?
25	What is the Kamchatka peninsula?	What is the Bering Strait?	What is continental?	What is Novosibirsk?	What is 1991?
BONUS	What is Communism Peak?	What is Lake Baikal (Baykal)?	What is chernozem?	What is Georgia?	What is communism?

7 Northern Eurasia
Topic: Occupations

	Agriculture	Forests, Fish, and Furs	Mining	Industries	Transportation and Trade
5	What is wheat?	What is lumber?	What is coal?	What is (iron and) steel?	What is the Volga River?
10	What is cotton?	What is transportation?	What is iron ore?	What is permafrost?	What are railroads?
15	What is grazing (or livestock raising)?	What are the Barents Sea and the White Sea?	What is the Kuznetsk Basin?	What is hydroelectricity?	What is St. Petersburg?
20	What is the potato?	What is hardwood?	What is oil?	What are textiles?	What is the Trans-Siberian Railroad?
25	What is the sugar beet?	What is the reindeer?	What is natural gas?	What are consumer goods?	What is Murmansk?
BONUS	What are sunflowers?	What is caviar?	What is manganese?	What is Ukraine?	What is an icebreaker?

8 | Asia
Topic: Physical Features

	Land	Water	Islands	Climate and Soil	Natural Resources
5	What are the Himalaya Mountains?	What is the Bay of Bengal?	What is Indonesia?	What are monsoons?	What is iron ore?
10	What is the Malay Peninsula?	What is the Yangtze River (Chang Jiang)?	What is Sri Lanka?	What is alluvial?	What is oil?
15	What is the Gobi Desert?	What is the Ganges River?	What are the Philippines?	What is the Japan Current (or Kuroshio)?	What is pollution?
20	What is Tibet (or the Tibetan Plateau)?	What is the Yellow River (Huang He)?	What is Honshu?	What is a typhoon?	What is coal?
25	What are the Pamirs?	What is the Indus River?	What is Java?	What is an earthquake?	What is natural gas?
BONUS	What is Taklimakan (or Takla Makan)?	What is the Mekong River?	What is New Guinea?	What are volcanoes?	What is tungsten?

8 | Asia
Topic: Occupations

	Agriculture— Crops	Agriculture— Methods	Fishing and Forestry	Manufacturing	More Income
5	What is rice?	What is a paddy?	What is seafood (or fish)?	What is silk?	What is fishing?
10	What is tea?	What are plantations?	What is teak?	What are people (or skilled labor)?	What is Malaysia?
15	What is cotton?	What is terracing?	What is fish-farming (or aquaculture)?	What is steel?	What is cottage industry?
20	What is rubber?	What is livestock?	What is reforestation?	What is Manchuria?	What are handicrafts?
25	What is jute?	What are pigs?	What is a sponge?	What is fertilizer?	What is Bangkok?
BONUS	What are soybeans?	What is slash and burn?	What is the (pearl) oyster?	What is electronic equipment?	What is copra?

8 Asia
Topic: People

	People and Problems	Transportation and Trade	Cities	Governments	Potpourri
5	What is (the People's Republic of) China?	What is foreign trade?	What is Tokyo?	What is Hong Kong?	What is Mount Everest?
10	What are rivers?	What is Singapore?	What is Shanghai?	What is Taiwan (or Formosa)?	What are (untapped) natural resources?
15	What is high birthrate?	What is the Association of Southeast Asian Nations (or ASEAN)?	What is Calcutta?	What are Laos, Vietnam, and Cambodia (Kampuchea)?	What is nuclear power?
20	What are cities (or shantytowns)?	What is the automobile?	What is Bombay?	What is Korea?	What is bamboo?
25	What is Tokyo?	What is the Khyber Pass?	What is Manila?	What are tax breaks?	What is Malaysia?
BONUS	What is Bangladesh?	What is a junk?	What is Beijing?	What is communism?	What is the Ganges River?

9 Southwest Asia and North Africa
Topic: Physical Features

	Land	Water	Climate and Soil	Natural Resources	Potpourri
5	What is the Arabian Peninsula?	What is the Nile River?	What is the Sahara Desert?	What is water?	What is the Strait of Gibraltar?
10	What is an oasis?	What is the Red Sea?	What is desert?	What is oil?	What is the Aswân High Dam?
15	What are the Atlas Mountains?	What is the Persian Gulf?	What is Mediterranean?	What is natural gas?	What is the Dead Sea?
20	What is the Sinai Peninsula?	What are the Tigris and Euphrates rivers?	What is the Nile River?	What are iron and coal?	What is a wadi?
25	What is the Hindu Kush?	What is the Shatt-al-Arab?	What is the Fertile Crescent (or Mesopotamia)?	What is cork?	What is Mauritania?
BONUS	What is Turkey?	What is the Dardanelles Strait?	What is Ar Rub' al Khali (or Empty Quarter)?	What is a sponge?	What is the Bosporus?

9 Southwest Asia and North Africa
Topic: Occupations

	Agriculture	Mining	Manufacturing	Other Income	New Ways
5	What is cotton?	What is Saudi Arabia?	What is the textile industry?	What is a sheep?	What is Turkey?
10	What are dates?	What is oil?	What is cottage industry?	What are Bedouins?	What is Israel?
15	What is wheat?	What is phosphate?	What is cement?	What is tourism?	What is a kibbutz?
20	What is citrus?	What is chromite?	What is potash?	What are pearls?	What is rice?
25	What is the olive?	What is iron ore?	What are (Persian) rugs?	What is caviar?	What is the Negev Desert?
BONUS	What is barley?	What is lapis lazuli?	What are assembly plants?	What are royalties?	What is afforestation?

9 Southwest Asia and North Africa
Topic: People

	Peoples	Transportation and Trade	Cities	Problems	Potpourri
5	What is Arab?	What is a camel?	What is Istanbul?	What is Israel?	What is farming?
10	What is Israel?	What is the Suez Canal?	What is Tehran (Teheran)?	What is irrigation?	What is Western Sahara?
15	What are Muslims?	What is the Organization of Petroleum Exporting Countries (OPEC)?	What is Cairo?	What is Afghanistan?	What is Baghdad?
20	What is Jerusalem?	What is a pipeline?	What is Tel Aviv-Yafo?	What is Cyprus?	What is Kuwait?
25	What is Mecca?	What is the Khyber Pass?	What is Riyadh?	What is salt?	What is the United Arab Emirates?
BONUS	What are Kurds?	What is the Arab League (League of Arab States)?	What is Beirut (Bayrūt)?	What is illiteracy?	What are bazaars?

10 Sub-Saharan Africa
Topic: Physical Features

	Land	Water	Climate	Natural Resources	Potpourri
5	What is the Great Rift Valley?	What is the Congo River?	What is rain forest?	What is leached?	What is the Sahara Desert?
10	What is a plateau?	What is the Niger River?	What is savanna?	What are waterfalls?	What are Victoria Falls?
15	What is Madagascar?	What is the Gulf of Guinea?	What is the Kalahari Desert?	What is Nigeria?	What is Mount Kilimanjaro?
20	What is Ethiopia?	What is Lake Victoria?	What is the high veld?	What is bauxite?	What are torrential rainstorms?
25	What are the Drakensburg Mountains?	What is Lake Tanganyika?	What is Mediterranean?	What is cobalt?	What is Djibouti?
BONUS	What is the Congo River Basin?	What is the Mozambique Channel?	What is the Sahel?	What is baobab?	What are elephants?

10 Sub-Saharan Africa
Topic: Occupations

	Agriculture	Forestry	Mining	Industries and Products	Potpourri
5	What is the peanut?	What is cacao?	What is gold?	What is the sheep?	What is subsistence?
10	What are nomads?	What is the oil palm?	What is copper?	What is oil?	What is the Kariba Dam?
15	What is cotton?	What is rubber?	What are gems (or jewelry)?	What is leather?	What is the Gebel Aulia Dam (Jabal Awliya Dam)?
20	What is coffee?	What is ebony?	What is Botswana?	What is sugar?	What is the Cape of Good Hope?
25	What is corn?	What is mahogany?	What is Liberia?	What is uranium?	What are safaris?
BONUS	What is millet?	What is the acacia?	What is manganese?	What is sisal?	What is the tsetse fly?

10 Sub-Saharan Africa
Topic: People

	Peoples	Transportation and Trade	Cities	Problems	Solutions
5	What is black?	What are rapids?	What is Cape Town?	What is South Africa?	What is independence?
10	What is Nigeria?	What are railroads?	What is Nairobi?	What is Democratic Republic of the Congo (or Congo-Kinshasa)?	What is ecotourism?
15	What is ethnic group (or tribe)?	What is the Congo River?	What is Johannesburg?	What are boundaries?	What is Kenya?
20	What are English and French?	What is landlocked?	What is Addis Ababa?	What are (experienced, honest) leaders?	What is wildlife?
25	What is Liberia?	What is Brazzaville?	What is Dakar?	What are ethnic (civil) wars?	What is Namibia?
BONUS	What is Bantu?	What is Khartoum?	What is Dar es Salaam?	What is foreign debt?	What is the Organization of African Unity (or OAU)?

11 Australia and Oceania
Topic: Physical Features

	Land	Water	Islands	Climate	Natural Resources
5	What is the Great Dividing Range?	What is the Murray River?	What is New Zealand?	What is desert?	What are artesian wells?
10	What is the Western Plateau?	What is the Darling River?	What is Tasmania?	What is a typhoon?	What is copra?
15	What are the Southern Alps?	What is the Tasman Sea?	What is Papua New Guinea?	What is tropical?	What is hydroelectricity?
20	What is Cape York Peninsula?	What is the Great Australian Bight?	What is coral?	What is marine?	What is uranium?
25	What is the Coastal Plain?	What is Lake Eyre?	What are high islands?	What is Mediterranean?	What is (geothermal) steam?
BONUS	What is the Great Artesian Basin?	What is a lagoon?	What is an atoll?	What are westerlies?	What is kiwi?

11 Australia and Oceania
Topic: Occupations

	Agriculture	Forestry and Fishing	Mining	Industries and Products	Other Income
5	What is wheat?	What is coconut palm?	What is coal?	What is wool?	What is agriculture?
10	What are beef cattle?	What is the Great Barrier Reef?	What are lead and zinc?	What are mutton and lamb?	What is tourism?
15	What is merino?	What is fishing?	What is bauxite?	What are dairy products?	What is foreign exchange?
20	What are dairy cattle?	What is New Zealand?	What is iron ore?	What is steel?	What are military bases?
25	What are apples?	What is the eucalyptus?	What is gold?	What is sugarcane?	What is Mount Isa?
BONUS	What are grapes?	What is the palm?	What is New Caledonia?	What are automobile plants?	What is Papua New Guinea?

11 Australia and Oceania
Topic: People

	Peoples	Transportation and Trade	Cities	Problems and Solutions	Potpourri
5	What is English?	What are refrigerator ships?	What is Sydney?	What is independence?	What is December?
10	What are Aborigines (or Aboriginals)?	What are airways?	What is Melbourne?	What is (Southeast) Asia?	What is the international date line?
15	What is Polynesia?	What is Japan?	What is Perth?	What is Melanesia?	What is the Tropic of Capricorn?
20	What is Micronesia?	What is England (or Great Britain)?	What is Brisbane?	What is New Zealand?	What is a playa?
25	What is Melanesia?	What is Fiji?	What is Auckland?	What is Micronesia (or Federated States of Micronesia)?	What is the outback?
BONUS	What is Oceania?	What is the railroad?	What is Canberra?	What is the (wild) rabbit?	What is a kangaroo?

12 Polar Regions
Topic: Natural Features

	Land and Water	Climate and Natural Resources	Occupations	Travel	Potpourri
5	What is the Bering Strait?	What is the tundra?	What is fishing?	What are icebergs?	What is an igloo?
10	What is the polar ice pack?	What is the midnight sun?	What is the whale?	What is a dogsled?	What are penguins?
15	What is the Antarctic Peninsula?	What is 66°30' (or Arctic and Antarctic Circles)?	What is scientific research?	What are airplanes?	What is Inuit (or Eskimo)?
20	What is the Ross?	What are high winds?	What is trapping?	What is an icebreaker?	What is the seal?
25	What are the Transantarctic Mountains?	What is plankton?	What are weather stations?	What is the Northeast Passage?	What is the Antarctic Treaty?
BONUS	What is the East Greenland Current?	What is coal?	What is oil?	What is the Northwest Passage?	What are krill?

13 World Perspectives
Natural Features

	Latitudes	Climate	Directions	Land	Water
5	What is 0° latitude (or equator)?	What are Antarctica and Europe?	What is Asia?	What is Australia?	What is the ocean?
10	What is Indonesia?	What is Antarctica?	What is south?	What is Panama?	What is the Suez Canal?
15	What is South America?	What is north?	What are the Philippines?	What is Great Britain?	What is the Caspian Sea?
20	What is the Tropic of Cancer (or 23½° north latitude)?	What is the Gulf Stream?	What is west?	What is Chile?	What is the Bay of Bengal?
25	What is aurora?	What is the jet stream?	What is east?	What is Norway?	What is the Bering Sea?
BONUS	What is the equinox?	What is windchill?	What is south?	What is the Mid-Atlantic Ridge (or Mid-Ocean Ridge)?	What is Lake Baikal (or Baykal)?

13 World Perspectives
Topic: People

	Population Changes	Environment	Problems	Solutions	Potpourri
5	What is demography?	What is the rain forest?	What is famine?	What is development?	What is indigenous?
10	What is migration?	What is acid rain?	What is national unity?	What is diversification?	What are fossil fuels?
15	What are service industries?	What is ozone?	What is desertification?	What is urban renewal?	What is the Ring of Fire?
20	What are refugees?	What is the greenhouse effect (or global warming)?	What is fish?	What is literacy?	What is Istanbul?
25	What is a megalopolis?	What is sustainable?	What is deforestation?	What is desalination?	What are seismic waves?
BONUS	What is Third World?	What is ecotourism?	What is poaching?	What is no-till?	What is gross national product (GNP)?

Share Your Bright Ideas

We want to hear from you!

Your name_____Date_____

School name_____

School address_____

City _____State _____Zip_____Phone number (_____)_____

Grade level(s) taught_____Subject area(s) taught_____

Where did you purchase this publication?_____

In what month do you purchase a majority of your supplements?_____

What moneys were used to purchase this product?

___School supplemental budget   ___Federal/state funding   ___Personal

Please "grade" this Walch publication in the following areas:

Quality of service you received when purchasing	A	B	C	D
Ease of use	A	B	C	D
Quality of content	A	B	C	D
Page layout	A	B	C	D
Organization of material	A	B	C	D
Suitability for grade level	A	B	C	D
Instructional value	A	B	C	D

COMMENTS:_____

_____

What specific supplemental materials would help you meet your current—or future—instructional needs?

_____

Have you used other Walch publications? If so, which ones?_____

May we use your comments in upcoming communications?   ___Yes   ___No

Please **FAX** this completed form to **888-991-5755**, or mail it to

Customer Service, Walch Publishing, P. O. Box 658, Portland, ME 04104-0658

We will send you a **FREE GIFT** in appreciation of your feedback. **THANK YOU!**